Communications
in Computer and Information Science 66

Yanwen Wu   Qi Luo (Eds.)

High Performance Networking, Computing, Communication Systems, and Mathematical Foundations

International Conferences, ICHCC 2009 – ICTMF 2009
Sanya, Hainan Island, China, December 13-14, 2009
Proceedings

 Springer

Volume Editors

Yanwen Wu
Intelligent Information Technology
Application Research Association
Hong Kong, China
E-mail: wyw1970_cn@yahoo.com.cn

Qi Luo
Wuhan Institute of Technology, China
Wuhan, Hubei, China
E-mail: witluo@ieee.org

Library of Congress Control Number: 2010922526

CR Subject Classification (1998): C.2, B.8, C.4, H.3.4, G.2, C.2.5

ISSN	1865-0929
ISBN-10	3-642-11617-5 Springer Berlin Heidelberg New York
ISBN-13	978-3-642-11617-9 Springer Berlin Heidelberg New York

springer.com

© Springer-Verlag Berlin Heidelberg 2010
Printed in Germany

Typesetting: Camera-ready by author, data conversion by Scientific Publishing Services, Chennai, India
Printed on acid-free paper   SPIN: 12828524   06/3180   5 4 3 2 1 0

Preface

The 2009 International Conference on High-Performance Networking, Computing and Communication Systems and the 2009 International Conference on Theoretical and Mathematical Foundations of Computer Science (ICHCC -ICTMF 2009) were held during December 13–14, 2009, in Sanya, Hainan Island, China. ICHCC -ICTMF 2009 was a comprehensive conference focused on the various aspects of advances in high-performance networking, computing, communication systems and mathematical foundations. The conferences provided a chance for academic and industry professionals to discuss recent progress in the area of high-performance networking, computing, communication systems and mathematical foundations.

The conferences were co-sponsored by the Intelligent Information Technology Application Research Association, Hong Kong and Wuhan Institute of Technology, China. The goal was to bring together researchers from academia and industry as well as practitioners to share ideas, problems and solutions relating to the multifaceted aspects of this area.

We received 60 submissions. Every paper was reviewed by three Program Committee members, and 15 were selected as regular papers for ICHCC -ICTMF 2009, representing a 25% acceptance rate for regular papers.

The participants of the conference had the chance to hear from renowned keynote speakers Jun Wang from The Chinese University of Hong Kong, Hong Kong and Chin-Chen Chang from Feng Chia University, Taiwan. We thank Springer, who enthusiastically supported our conference. Thanks also go to Leonie Kunz for her wonderful editorial assistance. We would also like to thank the Program Chairs, organization staff, and the members of the Program Committees for their hard work. Special thanks go to Springer CCIS. We hope that you enjoy reading the proceedings of ICHCC -ICTMF 2009.

December 2009   Qi Luo

Organization

The 2009 International Conference on High-Performance Networking, Computing and Communication Systems and the 2009 International Conference on Theoretical and Mathematical Foundations of Computer Science (ICHCC-ICTMF 2009) were co-sponsored by the Intelligent Information Technology Application Research Association, Hong Kong and Wuhan Institute of Technology, China

Honorary Chairs

Chin-Chen Chang	IEEE Fellow, Feng Chia University, Taiwan
Jun Wang	The Chinese University of Hong Kong, Hong Kong
Chris Price	Aberystwyth University, UK

General Co-chairs

Honghua Tan	Wuhan Institute of Technology, China
Qihai Zhou	Southwestern University of Finance and Economics, China

Program Committee Chairs

Zhu Min	Nanchang University, China
Xuemin Zhang	Beijing Normal University, China
Peide Liu	ShangDong Economic University, China
Xinqi Zheng	China University of Geosciences (Beijing), China

Publication Chairs

Luo Qi	Wuhan Institute of Technology, China
Yanwen Wu	Intelligent Information Technology Application Research Association, Hong Kong

Program Committee

Shao Xi	Nanjing University of Posts and Telecommunication, China
Xueming Zhang	Beijing Normal University, China
Peide Liu	ShangDong Economic University, China
Dariusz Krol	Wroclaw University of Technology, Poland
Jason J. Jung	Yeungnam University, Republic of Korea
Paul Davidsson	Blekinge Institute of Technology, Sweden

Cao Longbing   University of Technology Sydney, Australia
Huaifeng Zhang   University of Technology Sydney, Australia
Qian Yin   Beijing Normal University, China

Hosted by

Intelligent Information Technology Application Research Association, Hong Kong and Wuhan Institute of Technology, China

Supported by

International Journal of Intelligent Information Technology Application

Table of Contents

Research and Implementation of a Decoder of 1-Wire Bus

Yunping Wu[1,*], Weida Su[1], Conghui Chen[1], Yu Lu[1], Wangbiao Li[1], Yan Wu[1],
Lei Tang[1], and Shenzheng Cai[2]

[1] Department of Electronic and Information Engineering, Fujian Normal University. Fuzhou,
Fujian, 35007, P.R. China
[2] Faculty of Software, Fujian Normal University, Fuzhou 350007, P.R. China

Abstract. This paper designs a special schematic of signal separated circuit, which can distinguish and separate signals of 1-wire bus according to its sources, the decoder of 1-wire bus based on ARM and special schematic analysis signals' timing period runing through 1-wire, finally output procedure and details of 1-wire.

Keywords: Decoder, 1-wire, ARM.

1 Introduction

The 1-wire bus, a new type of bus technology, has many advantagements such as less lines and easy to use, only one line, through which it covers several fields, including control fields and data fields, which usually need many lines in traditional bus interfaces. Because of its many advantagements, the 1-wire bus is becoming widely used, more and more manufacturers introduced devices with 1-wire bus interface, such as DALLAS DS2401 silicon chip sequence and the DS18B20 temperature sensor[1][2], the DS80C400[3][4] with micro-Internet interface TINI (Tiny Internet Interfaces) etc..

Usually, there are master devices and slaver devices connected with 1-wire bus, only the masters can initiate a transaction. Devices detect signals of control and data depending on width of time of the high-low level whose time is strictly required[5][6]. For example: while the slaver is controlled to send its data, the master firstly output a high level to low level in more than 1us, and then, each bit of data from the slaver is output in every reading period of 15us[1].

It is quite difficult for traditional tools such as oscilloscopes and logic decoder to test and analysis signals runing on 1-wire bus besides a serial of wave, especially It can not be determined whether it is from the master or the slaver, and whether it represents data or controlling signal among a bound of wave. Therefore, A decoder of 1-wire bus which can distinguish signals according to its sources and analysis time period of each independent signal, can help us to know more details of 1-wire bus.

* The work was supported by Key project of Fujian Provincial Department of Science and Technology (2008H0022, 2009H0018), and the Bureau of Education of Fujian Province under Grant No. JB07080 and project of Fujian Normal University(2008100215).

Y. Wu and Q. Luo (Eds.): ICHCC-ICTMF 2009, CCIS 66, pp. 1–7, 2010.

2 Seperated Circuit Signal

Signal seperated circuit can not only distinguish the signals from the masters and the slaves but alse keep original system communicating normally.

2.1 Analysis

The schematic of signal separated circuit is shown in Figure1, in which A and B points connect respectively to the two ends of 1-wire, C and D points output signals distinguish ed by the circuit. C point outputs signal whose source is A, D point outputs signal whose source is B. Q1 and Q2 are NPN transistors, R1-R2-R3-R4 are pull-up resistors[7].

1-wire bus is simplex mode. That means only one device can control 1-wire at the same time. For example, in Figure1, B point is controlled when A point is outputing, on the contrary, A point is controlled while B point is outputing.

The analyse of circuit in Figure1 as follows :

i. Suppose that A point is controlling 1-wire and outputing high level, status of other points in Figure1 are listed:

$$V_{b1} = V_{e1} = V_{CC} \tag{1}$$

$$V_{b2} = V_{e2} = V_{CC} \tag{2}$$

Q1 and Q2 is closed , B and C and D are pulled to high level.

ii. Suppose that A point is controlling 1-wire and outputing low level, Q1-on, C point is becoming low level.

$$Vb1 = Ve1 + Vbe = Vbe \tag{3}$$

V_{be} is approximately 0.7V, B point outputs low level too ,because:

$$V_{e2} = V_{b1} = V_{e1} + V_{be} = V_{b2} + V_{be} > V_{b2} \tag{4}$$

Therefore, Q2 is closed, D remains high level.

Through the above analyse we can make a conclusions as following: what B and C points output is same as status of A point, what D point outputs is independent of A point. when B point controls 1-wire, the similar conclusions is achieved: what A and D points output is same as status of B point, what C point outputs is independent of B point.

To sum up, Signal separated circuit shown in the Fig 1 can distinguish and separate signals from 1-wire according to their sources. Both of B and C points output status of A point. Both of A and D points outputs status of B point. So follow-uping circuit can provide details of 1-wire by measuring time period of C and D points.

2.2 Simulation and Conclusions

The circuit shown in Fig 1 is simulated in Multisim9 software where resistance of R1-R2-R3-R4 are 4.7KΩ, Both Q1 and Q2 are 2N2102 transistors.

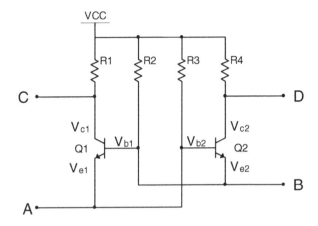

Fig. 1. The schematic of signal seperated circuit

The Fig 2 shows the waveform of B, C, D points when A point is controlling 1-wire input square of 5V@1KHz, output of B and C points whose amplitude is 4.3V keep pace with A point ouput.

The Fig 3 shows the waveform of A, C, D points when B point is controlling 1-wire input square of 5V@1KHz. output of A and D points whose amplitude is 4.3V keep pace with B point ouput.

The simulation results in Fig 2 and Fig 3 verify the conclusions of 2.1 that signal seperated circuit can distinguish and separate signals from C and D points. waveform of B and C points is the same as that of A point when A point is controlling 1-wire, waveform of A and D points is the same as that of B point when B point is controlling 1-wire. the delay of signal is very small, which means that circuit doesn't destruct the time period required by 1-wire bus and is able to keep the original system running normally.

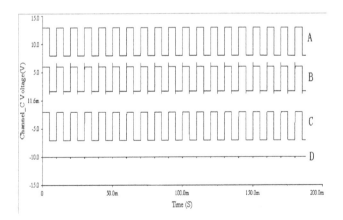

Fig. 2. The waveform of A,B,C and D points while A is controlling 1-wire

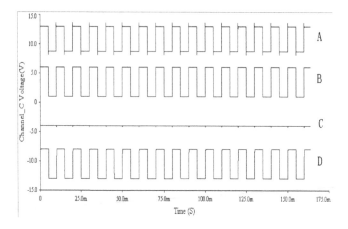

Fig. 3. The waveform of A,B,C and D points while B is controlling 1-wire

3   Design of Decoder of 1-Wire Bus

Decoder of 1-wire bus can be developed based on ARM embedded system and the signal seperated circuit. The dotted line in Fig 4 is a connections line of 1-wire bus in original system. ARM Microcontroller is LPC2114 based on ARM7TDMI-S core by PHILIPS, its main frequency to 60 MH[8]. Timing cycle of 1-wire bus is generally micro-second level, LPC2114's high perfomance can meet demand of the precision of measurement. In addition, it is convenient for measuring time of the signal cycle through LPC2114's capture functions. C and D outputs of signal seperated circuit connect respectively to 2 pins of LPC2114 which are capture mode and linking to Timer 0 and Timer 1, CPU will preserve the timers' value to the internal registers and generate an interruption when 2 pins change transition, and get time width of the transited period through timer values of two transited moment, and output these informations through the serial for debugging staff viewing or high-end software analysis time series.

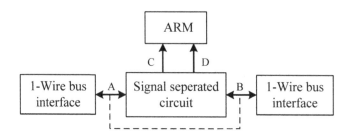

Fig. 4. The structure of a decoder of 1-wire bus

The code of initializing capture function of Timer 0 and Timer 1 as following:

// Open the capture function of P0.4, P0.10, and connect respectively to the capture function of Timer 0 and Timer 1.
PINSEL0 = (PINSEL0 & 0xFFCFFCFF) | 0x300300;
// Set the frequency of Timer 0 and Timer 1
T0PR = 0;
T1PR = 0;
// Open the capture function of Timer 0 and Timer 1, and have a interruption in the capture.
T0CCR = 0x30;
T1CCR = 0x06;
// Reset and start Timer 0 and Timer 1
T0TCR = 0x03;
T1TCR = 0x03;
The flowchart of ARM When ARM captures 2pins's change is shown in Fig 5.

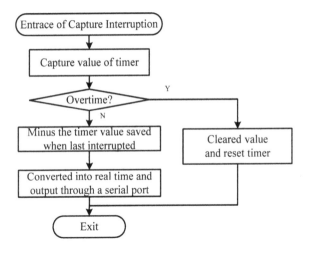

Fig. 5. The flowchart of the capture interruption

4 Application

One of applications of the decoder of 1-wire bus is to analysis procedure of comunicating through 1-wire bus between the master device and temperature sensor DS18B20. The original 1-wire is cut into 2 ends, one from master is connected to A point of the decoder and the other from DS18B20 is connected to B point. Fig 6 shows the waveform of signal on the 1-wire bus by the oscilloscope. Table 1 shows the part details of 1-wire by the decoder.

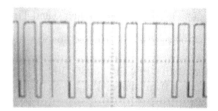

Fig. 6. The waveform of signal of 1-wire bus

Table 1. The part details of 1-wire by the decoder

No.	Point	High /Low	width (μS)	Remarks
1	C	Low	4	Master sends reading timing
2	D	Low	12	DS18B20 returns data '0'
3	C	High	76	Master releases 1-wire bus
4	D	High	70	DS18B20 releases wire bus
5	C	Low	4	Master sends reading timing
6	D	Low	12	DS18B20 returns data '0'
7	C	High	76	Master releases 1-wire bus
8	D	High	70	DS18B20 releases 1-wire bus
9	C	High	4	Master sends reading timing
10	C	High	76	Master releases 1-wire bus
11	C	Low	4	Master sends reading timing
12	D	High	70	DS18B20 returns data '1' and releases 1-wire bus when step9
13	D	Low	12	DS18B20 returns data '0' when step11

5   Conclusion

The signal separated circuit is simple and useful. It can effectively distinguish and separate signals at both ends of 1-wire bus and does minimum impact on the original communication, to keep the original system runing normally. The decoder based on ARM embedded system can analysis timing period of 1-wire bus, output alternating process of 1-Wire bus through the serial port, it is convenient for people to know much details of 1-wire.

References

1. Jun, M., Yanxiong, L., Changjun, L., et al.: 1-Wire Bus Technology Application in the Field of Test and Control. Chinese Journal of Scientific Instrumentation 22(S4), 254–256 (2001)
2. Xiaoyuan, W., Guang, C., Feng, L.: The Intelligent Temperature Control System with Long-distance Malfunction Diagnosis and Debugging Function. Chinese Journal of Scientific Instrumentation 23(S5), 167–170 (2002)
3. Bao-yu, Lianming, W., Xuefeng, X.: Study on the Technology of Embedded Internet Connection based on TINI Platform. Electronic Devices 30(2), 683–686 (2007)

4. Chuanjun, W., Yufeng, F., Dong, C.: The Net Meteorological Observatory based on TINI System. Astronomical Research & Technology 4(3), 288–295 (2007)
5. Guozhu, L.: Research of 1-Wire Bus Technology and Application. Journal of Xi'an University of Arts and Science: Natural Science Edition 9(2), 62–65 (2006)
6. Shengwen, L.: 1-Wire Device Bus Timing under C Compiler. Journal of Northeast China Institute of Electric Power 24(4), 82–85 (2004)
7. Zhichuan, C.: 1-Wire Bus Technology Applications Exchange of Experience. Guangdong Public Security Technology (3), 81–84 (2006)
8. LPC2114 DataSheet [CDROM] NXP Semiconductors

Design of the Energy and Distance Based Clustering Protocol in Wireless Sensor Network

Qingzhang Chen, Dina Fang, and Zhengli Wang

College of Computer, Zhejiang University of Technology, Hangzhou, China
qzchen@zjut.edu.cn

Abstract. For the typical clustering protocol LEACH, cluster head is selected randomly and the clustering distribution is unreasonable which both make the network energy conservation unsatisfactory. A routing protocol named EDBCP (Energy and Distance Based Clustering Protocol in Wireless Sensor Network) was proposed in this paper. It chooses and distributes the cluster heads more reasonability by considering the remaining energy, the distance between note and the base station borrowing ideas of roulette wheel selection of Genetic Algorithm and setting a distance threshold between cluster heads. Compared with LEACH, ECBCP has the greater capability on saving energy and prolong the network's lifetime.

Keywords: Wireless Sensor Network, energy, distance, cluster routing protocol, network survival time.

1 Introduction

WSN is a kind of wireless network made up of many sensor nodes. It can be used for sensing, collecting and processing information in an area which is disposed by collected nodes in real time, such as temperature, light strength, noise and other physical phenomenon and collect and deal with this information, then it send to observer in wireless.

Routing technology is a core of WSN. It has own characteristics, such as energy of the node could not be added. Each node has restriction on power in wireless sensor network. So Routing Protocol in wireless sensor network is designed to meet the application requirements minimizing the cost of network, promoting the load balancing and improving the survival time of the network.

So far, dozens of routing protocols of wireless sensor network have applied in different occasions. The saving of energy also has the different effect [1].This paper proposes an energy and distance based clustering protocol-EDBCP which improves the random selection defects of cluster heads choosing in LEACH algorithm. Borrowed ideas from roulette wheel selection of Genetic Algorithm to choose the cluster heads more reasonable. Energy consumption is much balance. Set a distance threshold between cluster heads, so every two cluster heads should maintain a certain distance. Then the distributing of cluster heads made more uniformity, hierarchical of network more effectively. After designating the cluster heads, the base station will inform simple nodes to join in a proper cluster, guaranteeing that the distance between a

Y. Wu and Q. Luo (Eds.): ICHCC-ICTMF 2009, CCIS 66, pp. 8–15, 2010.

simple node and the base station through the cluster head as short as possible. For some nodes, the consumption of energy which sends the information directly to the base station is much lower than to its cluster head. In that case ,the node will communicating with the base station directly to avoid wasting energy in the entire network and improve efficiency. At last, compared with LEACH, EDBCP prolongs the network's lifetime and saving energy more effectively.

2 Design of EDBCP

2.1 EDBCP Model

2.1.1 Energy Consumption Model of EDBCP

The energy consumption formula of EDBCP is the first order radio model [2]. The energy consumption of sensor nodes includes energy-aware, processing and transmission and so on. The major consumption of energy in radio energy consumption model is sensor nodes' communication. So this article considers the information transmission part most. The related formulas of this model are as follows:

Sensor nodes send k bit data consume energy:

$$E_{tr}(k,d) = E_{elec}(k) + E_{amp}(k,d) = \begin{cases} kE_{elec} + k\varepsilon_{fs}d^2 & d < d_0 \\ kE_{elec} + k\varepsilon_{amp}d^4 & d \geq d_0 \end{cases} \tag{1}$$

Sensor nodes receive k bit data consume energy:

$$E_{rx}(k) = kE_{elec} \tag{2}$$

In the formulas, E_{elec} is the energy consumption by transceiver of wireless circuit. E_{amp} is the energy consumption of the amplifier. Parameters ε_{fs} and ε_{amp} depend on the transmission amplifier model, d is the distance between sender and receiver, k means the number of transmission bits. $d < d_0$ and $d > d_0$, the two kinds of energy attenuation model named free space and multi-channel attenuation model.

2.1.2 EDBCP Timing Model

In accordance with the timing model in figure 1, the running of network is as follows: First is the network initialization phase T_{init}, followed by operational phase. In the operational phase, the first round is different in timing and other running rounds have the same time. The time of first round is expressed: T_{round1}. Other time of rest rounds are denoted by T_{round1}. In the first round time: T_{round1}, it is divided into four phases: the communications between all the sensor nodes and the base station; the selection of clusters and the formation of clusters; the communications in cluster; the communication between cluster heads and base stations. They are Expressed: T_{to_BS}, $T_{cluster}$, T_{to_CH} and $T_{C\_to\_BS}$. The rest rounds time is: T_{round}, they are also divided into four phases: the communications that last round's cluster heads send their energy information to the base station; the selection of clusters and the formation of clusters; the communications in cluster; the communication between cluster heads and base stations. They are

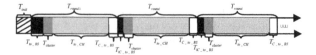

Fig. 1. EDBCP timing model

Expressed: $T_{IC\_to\_BS}$, $T_{cluster}$, T_{to_CH} and $T_{C\_to\_BS}$. The first round time is different with other rounds in the first phase. Because in the final phase of the first round time: $T_{C\_to\_BS}$, each simple sensor node has sent their collected data and remained energy information to the base station through cluster heads. In the followed round's beginning phase the simple sensor nodes needn't send their energy message to base station. The cluster heads sent so many messages to base station in the last round that much energy lost. So in the followed new round time: $T_{IC\_to\_BS}$, the cluster heads of last round should send their remained energy message to the base station. Then base station can running centralized computing.

2.2 Initialization of Network in EDBCP

In the initialization phase of the network, all the sensor nodes are the random distributed in the surveillance region. Base station is located in the central of the region which is coordinated as $S_{BS}(X_{BS}, Y_{BS})$, $X_{BS} = X/2$, $Y_{BS} = Y/2$. The total number of sensor nodes is n. The coordinate of each node is $S_i(x_i, y_i)$ $i = 1,2,3......n$, and all the sensor nodes have the same initial energy E_0. Assuming every node can be specified located manually through GPS or other methods to obtain its location information. The base station broadcasts its location information in the region.

After establishing the network, in EDBCP, sensor nodes transmit the energy, distance and any other messages. So the data transmission among nodes includes: energy message, named EM; the energy information and distance message, named EDM; single node Message, named SM; data message is named DM. EDM, EM and SM data are also assumed to be constant like LEACH.

2.3 Node Sending Message to Base Station

2.3.1 The Network Running in First Round Time

After the initial establishment of the network,   EDBCP immediately enters into the first round time. Each node should send its own remainder energy and location information to the base station. Energy consumption is free space model. The distance between node i and base station is:

$$d_{to\_BS}(i) = \sqrt{(x_i - X_{BS})^2 + (y_i - Y_{BS})^2} \qquad i \in 1,2,3......n \qquad (3)$$

Energy consumption of each node sending energy and location information to the base station is:

$$E_{to\_BS} = EDM * E_{elec} + EDM * \varepsilon_{fs} d_{to\_BS}^2 \qquad (4)$$

After all sensor nodes sent message to the base station, nodes are in the state of monitoring the channel waiting for receiving the message sent from the base station.

2.3.2 The Network Running of the Rest Cycle

As mentioned in the 2.1.2, it is different in the first phase between first round and rest rounds. In the $T_{IC\_to\_BS}$ of rest rounds, only the cluster heads of the last round send their energy information to the base station. The energy consumption equation of each cluster head when sending information is as follows:

$$E_{to\_BS} = EM * E_{elec} + EM * \varepsilon_{fs} d_{to\_BS}^2 \qquad (5)$$

2.4 The Method of Grouping Cluster in EDBCP

When receiving the energy and location information about all nodes, base station would carry out the selection of cluster heads, the algorithm of simple nodes assigned into cluster and so on. After calculating, base station sends the information of all sensor nodes to all nodes, such as node belongs to which cluster head, the TDMA slot assigned and so on.

2.4.1 The Selection of Cluster Heads in EDBCP

In LEACH, the choice of p value is very important. There are lots of studies on the p selection [3] [4] [5] [6]. The optimization number of cluster heads is:

$$c_{opt} = \sqrt{\frac{n}{2\pi}} \cdot \frac{2}{0.765} \qquad (6)$$

n is the number of sensor nodes which the remainder energy is greater than zero in the current round.

There are two kinds of cluster heads' selection in EDBCP. One is that some nodes can be chosen as the cluster heads by cluster heads selection algorithm. The other is some nodes could not be chosen as cluster heads under certain conditions.

1) Nodes are selected as cluster heads

In EDBCP, it wants to achieve the load balance of the network considering the distance between sensor node and base station and the node's remained energy. So EDBCP gets ideas from roulette wheel selection of Genetic Algorithm to choose the cluster heads. Get sensor node's energy and square of the distance between node and the base station as the individual fitness ratio: f_i

$$f_i = \frac{E_{current}}{d_{to\_BS} * d_{to\_BS}} \qquad (7)$$

$E_{current}$ is the current energy of sensor node. d_{to_BS} is the distance between the node i and the base station. The probability of sensor node selected as the cluster heads is:

$$P_i = \frac{f_i}{\sum_{j=1}^{n} f_j} \qquad (8)$$

The sum of all the nodes' P_i is 1. The number of sensor nodes is n. Roulette is divided into n regions in accordance with its fitness value by the way in clockwise to be distributed. Each region occupies P_i proportion. Generate a number between 0 and 1 randomly. The node is selected as the cluster head which the number denotes. Recycled to produce a random number to select next cluster head which have not been chose before. And go on until the number of cluster heads reach the optimal number: $c_{opt}(r)$. This method of selecting cluster heads made the nodes that with higher energy and close to the base station become cluster heads more likely. But not only nearer base station node become the cluster heads. Each live node can be the cluster head. Only each node has different probability.

2) Nodes could not be selected as cluster heads
In the process of selecting cluster heads, the distance between some cluster heads may be very close sometimes. The uneven distribution in cluster heads may make the waste. So in EDBCP, it sets a threshold of distance between cluster heads:

$$Th = \frac{\sqrt{X * Y}}{\theta} \tag{9}$$

X and Y are the region's length and width, θ is a constant (when length and width are 100 meters, the optimization value of θ is 20 from a series tests). The distance between cluster heads should be larger than Th:

$$Dis\{C_i C_j\} > Th \qquad i, j \in 1, 2 \cdots\cdots C_{opt}(r) \tag{10}$$

2.4.2 Cluster Building
After selecting the cluster heads, base station have to choose some nodes to join the cluster. Generally speaking, the sensor nodes in the network have to belong to a cluster [7]. Node which enters the cluster is also divided into two kinds: general nodes and special nodes.

(1) General nodes
Because of the node's power lost in transmission and the square of the distance in transmitting is directly proportional. To lower the transmission distance is to reduce the energy consumption in transmission. The idea of EDBCP is to reduce the total energy consumption of simple node sending information to the base station through cluster head. That is to say, the transmission path from a simple node to the base station through the cluster head will be as short as the straight-line distance of two-dimensional plane.

EDBCP makes reference to the MTE [7]. Node will choose cluster heads one by one. $d^2_{to\_ch\_BS}$ is the sum of square of distance from node to cluster head and the distance from the cluster head to base stations:

$$d^2_{to\_ch\_BS} = d^2_{to\_ch}(cj) + d^2_{to\_BS}(cj) \qquad cj \in I \tag{11}$$

cj in the formula is the cluster head's serial number. d_{to_ch} is the distance from node to its cluster head cj. d_{to_BS} is the distance from cluster head cj to base station. I is the aggregate of all the cluster heads. The square sum of the distance from node to base station is.

$$d^2_{to\_ch}(c_j) + d^2_{to\_BS}(c_j) = ((x - x_{cj})^2 + (y - y_{cj})^2) + ((x_{cj} - x_{BS})^2 + (y_{cj} - y_{BS})^2) \tag{12}$$

According to the formula, the base station calculates each node's minimum. The shortest square sum of distance from cluster head to base station is:

$$dis\_to\_bs^2 = \min\{d^2_{to\_ch}(cj) + d^2_{to\_BS}(cj)\} \tag{13}$$

Its minimum value is the cluster head that the node to join. The energy consumption of the cluster head transmit node's message to the base station is:

$$E = DM * E_{elec} + DM * \varepsilon_{fs} d^2_{to\_BS}(cj) \tag{14}$$

(2) Special nodes
There are always some nodes that are nearer base station. As shown in Figure 3:

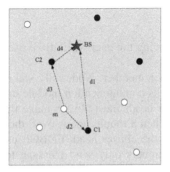

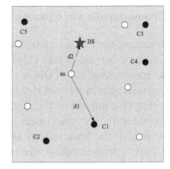

Fig. 2. The case of simple node   **Fig. 3.** The case of special node

$d^2_1 > d^2_2$. That means the energy consumption of sn node sending information directly to base station is fewer than to the cluster head C1. So in EDBCP, the base station distributes TDMA slot in which each special node send data to the base station directly with CSMA MAC. Or the special node will close the communication module to stop monitor channel when out of TDMA slot. The energy consumption that send message to base directly are as follows:

$$E_{to\_BS\_Data} = DM * E_{elec} + DM * \varepsilon_{fs} * d^2_{to\_BS} \tag{15}$$

3   The Process of Communication in EDBCP

This section will further describe the communication in the entire network in detail. Figure 1 has given the timing model of EDBCP. Now it introduces as follows: the communication of base station and sensor nodes, communication in cluster and among the clusters.

3.1 The Communication between Sensor Node and Base Station

The energy of base stations is unlimited. The communication process between base station and sensor nodes is as follows:

1. After a short time of T_{init}, the base stations broadcast synchronization signal in the initial phase of the network and then network gets into the stage of the establishment.

2. Every sensor node sends a short message to the base station with CSMA MAC protocol including the remained energy and location of the nodes.

3. Base station uses CSMA MAC protocol to broadcast the messages of all the sensor nodes after chosen the cluster heads and built clusters. The broadcast messages include node ID, CDMA spread-spectrum code of clusters, and TDMA slot of each node and so on.

4. The special node data: as they are near the base station, they distributed their TDMA slot by base station. Then nodes use CSMA MAC protocol in their own slots to send data monitored by sensor nodes.

3.2 Communication in Clusters

For promoting the efficiency of transmission and saving the energy, EDBCP uses the mechanism of dormancy and TDMA like LEACH.

Base station allocates TDMA slot for each cluster member. Node only wakes up and transmits data in its own slot time. Node closes its communication module to stop monitoring channel out of the slot time. The cluster heads must always make the receiver open to receive the cluster members' data. When a round time is over, the data will be sent to the base station after processed by the cluster head. EDBCP used a single jump communication that means cluster members only send message to the own cluster head.

In the sensor network, the same object may be monitor by several sensor nodes or several nodes monitor the similar environment. The report data of these nodes will be same or similar. The cluster heads will fuse data to get rid of the redundancy information which minimizes the amount of data before sending to the base station.

3.3 Communication among Clusters in EDBCP

The two neighbor clusters have signal interference caused by the signal coverage. In order to avoid frequency interference among cluster, it uses CDMA mechanism in cluster. The base station allotted CDMA code of each cluster and broadcast the current CDMA code with TDMA time slot. After receiving the message, they sent message to their cluster heads according to the CDMA code in their TDMA time slot. So in this way, when communicating in the cluster, signal of other clusters will be filtered to avoid collision [2].

4   Compared with LEACH and Conclusion

From the mass experiment, under the same condition and parameter, the deaths nodes on the number in EDBCP are basically the least in each round. The failure time of 1%

dead nodes and 50% dead nodes are all much higher than that in LEACH which delivered directly in plane. So the conclusion is that EDBCP not only save nodes' energy but extend the life time of network. In the case of different nodes density lifetime and stability of network in EDBCP is still better than in LEACH. Change Nodes' density, EDBCP remains in force. With the reduction of data, the running time of network increased. This means EDBCP adapt to the change of data capacity and also illustrates the necessary of application of data fusion technology.

This paper brought forward a protocol of EDBCP which improves the LEACH that did not consider the cluster heads selection in energy heterogeneous. It took the remained energy and distance as parameters, borrowed ideas from roulette wheel selection of Genetic Algorithm to choose the cluster heads, controlled the cluster heads' distance at the same time to Make distribution of cluster heads more uniformity and reasonable. After confirming the cluster heads, the base station will inform simple nodes to join in a proper cluster, guaranteeing that the distance between a simple node and the base station through the cluster head as short as possible. For some special nodes, they communicate with base station directly. Avoid the energy waste of whole network. Using TDMA and CDMA technology in and between the clusters further improve the energy efficiency of the protocol. Compared with LEACH, and EDBCP, the result is that EDBCP had great improved in load balancing, and can extend the survival time of the wireless network effectively.

Acknowledgements

The Project Supported by Zhejiang Provincial Major Project of China (**2007C13064**).

References

1. Jamal, N., Kamal, A.E.: Routing Techniques in Wireless Sensor Networks: A Survey, pp. 1696–1705. IEEE Computer Society, San Francisco (2002)
2. Heinzeiman, W., Chandrakasan, A., Balakrishnan, H.: Energy-efficient communication protocol for wireless micro sensor networks. In: Proceedings of the 33rd Hawaii International Conference on System Sciences, Maui, pp. 3005–3014. IEEE Computer Society, Maui (2000)
3. Bandyopadhyay, S., Coyle, E.J.: Minimizing communication costs in hierarchically clustered networks of wireless sensors. In: Wireless Communications and Networking, WCNC 2003. 2003 IEEE, March 16-20, vol. 2(2), pp. 1274–1279 (2003)
4. Heinzelman, W.R., et al.: An application-specific protocol architecture for wireless microsensor networks. IEEE Trans. on Wireless Communications 1(4), 660–670 (2002)
5. Bandyopadhyay, S., et al.: An energy-efficient hierarchical clustering algorithm for wireless sensor networks. In: Proc. of IEEE INFOCOM, pp. 1713–1723. IEEE Computer Society, San Francisco (2003)
6. Smaragdakis, G., Matta, I., Bestavros, A.: SEP: A stable election protocol for clustered heterogeneous wireless sensor networks. In: Proc. of the Int'l Workshop on SANPA 2004 (2004)
7. Ye, M., Li, C., Chen, G., Wu, J.: EECS: An energy efficient cluster scheme in wireless sensor networks. In: Dahlberg, T., Oliver, R., Sen, A., Xue, G.L. (eds.) Proc. of the IEEE IPCCC 2005, pp. 535–540. IEEE Press, New York (2005)

The Design of NTCIP C2F Model System for ITS

Qian Yan and Yin Wen-qing

Institute Of Technology, Nanjing Agricultural
University, Nanjing 210031, P.R. China
qianyan@njau.edu.cn

Abstract. NTCIP (National Transportation Communications for ITS Protocol)
is the standard communication protocol for Intelligent Transportation Systems
(ITS) of the data transmission between the traffic control devices and ITS system. NTCIP protocol for ITS is introduced in this paper, The design methods of
C2F (Center-to-Field) which is the main aspect of NTCIP are given. Through
the application in a program, a common model of the Center-to-Field design is
proposed. All the related modules can be used in an actual program, and the
details of the model are analyzed for reference.

Keywords: NTCIP, ITS, Center-to-Field model.

1 Introduction

Nowadays, intelligent traffic control is applied in more and more fields. ITS (Intelligent Traffic System) give the methods to build highway in our country include national information collection, report, analysis, and announcement system. The traffic
management system is designed to implement the traffic monitor, traffic signal, traffic
guiding technology on national highways and national main ways.

National Transportation Communications for ITS Protocol (NTCIP) is a standard
protocol which is created by Federal Highway Administration (FHWA), Institute of
Transportation Engineers (ITE), American Association of State Highway and Transportation Officials (AASHTO) and National.

Electrical Manufacturers Association (NEMA) who joint committee. NTCIP is a
set of protocols, which include Center-to-Center communication and Center-to-Field
communication. The former is the information communication between the computers
of administrator center and computers of all the subnet system, while the later is the
information communication between facilities and vehicles in the roads.

2 NTCIP Framework

NTCIP is composed by five layers, which include Information Level, Application
Level, Transport Level, Sub-network Level and Plant Level. The structure of these
levels is shown in Figure1.

Y. Wu and Q. Luo (Eds.): ICHCC-ICTMF 2009, CCIS 66, pp. 16–21, 2010.

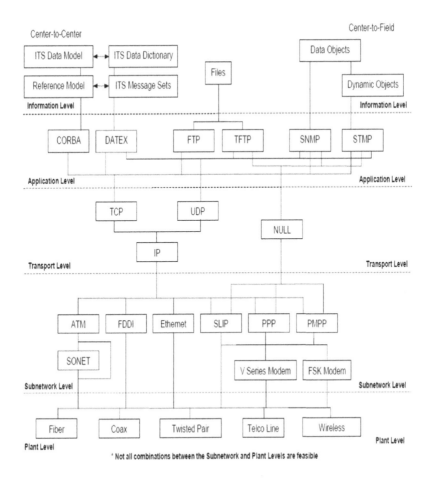

Fig. 1. NTCIP framework

Information level offers the standard formatting definition and code of data, object and information which application program processes.

Application level offers the structure of information package and the standard of the information management.

Transport level offers the standard of package, combinatory and routing etc.

Sub-network level offers the standard of object interface such as Modem , NIC, CSU/DSU etc.

Plant Level includes the communication media such as Twist Pair, coaxial cable, Fiber and wireless communication etc.

3   C2F Model System Framework

C2F (Center-to-Field) Model System is surrounded with a project framework model which is based on ITS NTCIP.This project of Intelligent Traffic System is based on

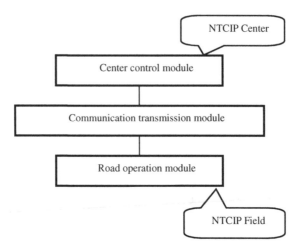

Fig. 2. System framework on C2F

three modules: center control module, communication transmission module and road operation module. The framework of the system is concerned in Figure 2.

3.1 Center Control Module

Center control module is composed with center control computer and field control computer. The principle functions are fault detection and management of equipment and software; analysis and implement of traffic control information; data management of traffic flow and the management of the interface; showing the real-time condition and information of traffic; implement of priority to public transportation; management of traffic accident etc (as Figure3).

3.2 Communication Transmission Module

Communication transmission module is composed with communication control equipments (including switch, optical transmitter and receiver, serial port server etc.), Fiber, Internet. The principle function is the link of all the communicating data of all the system, which is like a bridge.

3.3 Road Operation Module

Road operation module is composed with equipments which including traffic signal controllers, vehicle testing machines, traffic lights and the screens of countdown. The principle functions are the identification and pretreatment of traffic signal, optimization calculation of circle and green single rate, implement of online control and single control, output of light control single, control of the screen of countdown in three colors in frequency conversion circuit and monitor the fault of the equipments above(as Figure 4).

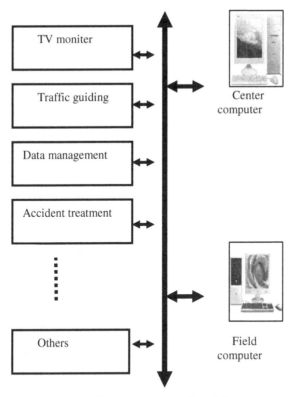

Fig. 3. Center control module

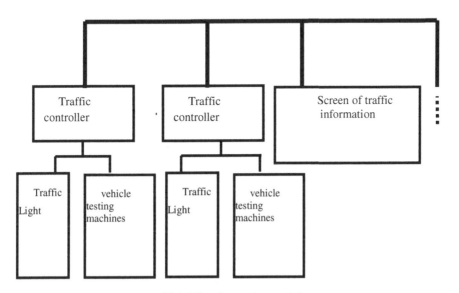

Fig. 4. Road operation module

4 Key Technology of System

NTCIP Center-to-Field(C2F) communication of ITs, which involves Simple Transportation Management Protocol(STMP) which is application level protocol, subnet-
work level protocol Point-to-MultiPoint Protocol(PMPP), and the format standards in
information level , which is called Object.

STMP protocol supports low bandwidth link, so it can runs on T2/NULL, but if the
bandwidth is enough, it can also runs on UDP/IP or TCP/IP network. To design an
actual system, the common protocol units are STMP(application layer), T2(translate
layer) and PMPP(subnet layer), which is called STMP over T2 over PMPP.

4.1 STMP Protocol

STMP protocol implement management functions of C2F communication, which lies
in the agent of data transport.

Compare with other protocols on the same level, STMP can economize more
bandwidth by effective coding mechanism to support Dynamic Composite Object. So
STMP is a kind of choice which has elastic properties.

The comparative between STMP and SNMP is shown in Table1.

Table 1. Comparative between STMP and SNMP

Function	SNMP	STMP
All the basic data type	Yes	Yes
Bandwidth usage rate	Low	High
Routing and dialing	Yes	Yes
Information Set	Few	More
Operation complexity	Low	High

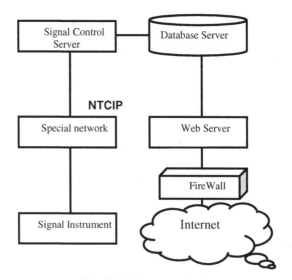

Fig. 5. Model system design

4.2 System Design

Center-to-Field communication is between center computer and field equipment.

Key point in center computer is the Web operation process, dynamic Web page design technology based on XML.

In field computer, embedded development technology is used, which use embedded Linux operation system, serial port to support communication.

The main design of modules is shown in Figure 5.

5 Conclusion

In this paper, a model of C2F theory which is in ITS NTCIP protocol is designed. Firstly, the overview of NTCIP protocol is given, which includes the definitions of five levels. Secondly, an actual program system is shown to guide the design, this system is based on NTCIP Center-to-Field theory. Finally, key technology of the design of the system model is concerned.

References

1. Zhang, B.-h., Xiao, Y.-y.: The Abstract of National Traffic Communication Protocol NTCIP (1). China Traffic Information (2004)
2. Zhang, B.-h., Xiao, Y.-y.: The Abstract of National Traffic Communication Protocol NTCIP (2). China Traffic Information (2004)
3. NTCIP, The National Transportation On Communications for ITS Protocol Guide
4. AASHTO\ITE \NEMA.NTCIP 1103 v01.26a: National Transportation Communications for ITS Protocol Transportation Management Protocols. A Recommended Standard of the Joint Committee on the NTCIP (June 2005)
5. AASHTO\ITE \NEMA. NTCIP 1204 v03.03c: Environmental Sensor Station Interface Standard – Version 03. National Transportation Communications for ITS Protocol (June 2007)
6. The Joint Committee on the NTCIP. The NTCIP Guide (NTCIP 9001) version 3[G/OL], http://www.Ntcip.org 2002-10
7. Demetsky, M.J., Brian Park, B., Venkatanarayana, R.: A Research Project Report For the Virginia Department of Transportation[G/OL], http://Nets.Virginia.edu/docs/UVACI-13-15-52.pdf 2001-11
8. The Joint Committee on the NITCIP.NITCIPII Guide (NTCIP 9001)~,elision 3[EB/OL] [2006-08-11], http://www.Ntcip.org
9. NTCIP 1202. 0 object definitions for a actuated Traffic Signal Controller Units (2002)

Scheduling Algorithm for Vehicle to Road-Side Data Distribution

Mohammad Shahverdy, Mahmood Fathy, and Saleh Yousefi

Young Researchers Clup, Azad University- Tafresh Branch, Tafresh, Iran
Computer Engineering Faculty, Iran University of Science and Technology
Computer Engineering Department, Urmia University, Urmia, Iran
shahverdi@iautb.ac.ir, mahfathy@iust.ac.ir,
s.yousefi@urmia.ac.ir

Abstract. We investigate the problem of scheduling of file distribution from Road-Side Units (RSUs) to vehicles in an urban environment. We suppose one RSU exist in each cross-road and all RSUs are connected to each other through a high speed backbone. Since the number of vehicles as well as the number and size of files are potentially large, scalable scheduling is a challenge. We first chop each file into several segments. A vehicle may not be able to finish downloading all segments from a RSU, thus, our proposed algorithm allows it to continue its download from the next RSUs. We propose a scheduling algorithm in which each RSU implements two separate queues for: 1) download requests from the scratch and 2) resumed downloads. The data segment for distribution is chosen from aforementioned queues based on some scheduling policies. Simulation result shows that the proposed approach offer desirable performance and scalability.

Keywords: Vehicular ad hoc network (VANET), Roadside unit (RSU), Scheduling scheme, queue.

1 Introduction

Due to recent surge of interest to vehicular networks, it is expected that data access from Road-Side Units (RSUs) becomes crucial in the near future. Recently, vehicle-roadside data access has received considerable attention [1, 2, 3, and 4]. With Road Side Unit (RSU) such as 802.11 access point, vehicles can access data stored in the RSU or even access the Internet through these RSUs. The Federal Communications Commission (FCC), realizing the problem of traffic fatalities in the US dedicated 75 MHz of the frequency spectrum in the range 5.850 to 5.925 GHz to be used for vehicle to vehicle and vehicle to roadside communication. The potential applications for the standard which is called Dedicated Short Range Communication (DSRC) include safety applications and comfort applications [5].

This paper studies a wide group of comfort applications which need vehicle-to-roadside unit communications. In this case RSUs can act as buffer point between vehicles or act as a router for vehicles to access the internet. Moreover, the RSUs may

Y. Wu and Q. Luo (Eds.): ICHCC-ICTMF 2009, CCIS 66, pp. 22–30, 2010.
© Springer-Verlag Berlin Heidelberg 2010

act as server and thus provide various type of information to vehicles on roads. The following instances are some example for RSU applications:

1. WEB Applications: The passengers can connect to the internet and use of it, such as checking Emails, browsing pages or other web applications.

2. Real Time Traffic: Vehicles can report real time traffic observations to RSUs. The traffic data then can be transmitted to a traffic center. The result of traffic data analysis then can be accessible to vehicles moving across each RSU.

3. Digital Map Downloading: When vehicles driving to a new area, they may hope to update map data locally for travel guidance such as changing unilateral or deadlock roads.

Vehicles are moving and they only stay in the RSU area for a short period of time. When number of requests is increased, an important challenge is to implement a suitable scheduling algorithm that serves more requests as possible. Furthermore, reducing delay of downloading is also another goal of scheduling algorithm. We suppose one RSU exist in each cross-road and all RSUs are connected to each other through a high speed backbone. Since the number of vehicles as well as the number and size of files are potentially large, scalable scheduling is a challenge. We first chop each file into several segments. A vehicle may not be able to finish its download from a RSU, thus, our proposed algorithm allows it to continue its download from the next RSUs. We propose a scheduling algorithm in which each RSU implements two separate queues for: 1) download requests from the scratch and 2) download requests which are resumed. The data for distribution is chosen from aforementioned queues based on some scheduling policies. Simulation result shows that the proposed approach outperforms other previously scheduling approached in sense of scalability.

The rest of this paper is organized as follows. In section 2 we first introduce related works. In section 3 system model and assumptions are discussed. In section 4 we highlight the importance of being able to resume downloads in each RSU. In section 5 we present a method for improve scheduling scheme called two queue method and compose this with D*S scheduling scheme. Section 6 evaluates the performance of the proposed scheme by means of simulation. Finally we conclude the paper in Section 7.

2 Related Work

Recently vehicle to RSU communications are studied in a few works. In most previous work, requests do not have deadline constraints (e.g., average/worst-case waiting time and throughput) [6, 7, 8, 9, and 10]. However, some works take into account some time deadline which should be met by the RSU [11]. If the RSU can not met the time deadline for some requests and vehicles move out from the RSU area, their request will be dropped completely. In this case, vehicles start over the whole request from another RSU even if a part of data has been already downloaded. This issue leads to weak throughput and delay performance. In our paper we address this challenge.

We implemented our schemes in application layer while other schemes work further on other layer [1, 2, 3, and 4]. The goal of [1] is to prove that WLAN technology is capable of enabling Drive-thru Internet access, wherein, the authors focus on plain

WLAN connectivity and transport protocol behavior. However, they briefly address implications on applications. In [4], author work on MAC layer to improve vehicular opportunistic access from roadside access point. For this goal, wireless conditions in the vicinity of a roadside access point are predictable, and by exploiting this information, vehicular opportunistic access can be greatly improved Although some paper [11], deal their scheduling schemes on application but they do not take into account large scale of roads and more importantly they do not differ between vehicles that request files form scratch and those request the file form remaining part. As it was shown these approaches show low throughput if the number of vehicles and the size of files increases.

Fig. 1. The System Model Architecture

3   System Model and Assumptions

We consider a Manhattan with nine cross-road that is all street plumb together. In each cross-road one RSU exists [12] (Figure 1 shows a part of our scenario). A number of vehicles retrieve their data from the RSU when they are in RSU's coverage range. The RSU (server) maintains a service cycle, which is non-preemptive; i.e., one service can not be interrupted until it is finished. All vehicles can send request to the RSU if they tend to access the data. Each request is characterized by a 5-tuple: <v-id, d-id, w-RSU, s-rec, deadline >, where v-id is the identifier of the vehicle, d-id is the identifier of the requested data item ,w-RSU identifies of the RSU form which the vehicle has come from its area, s-rec is identifier of data size that is received until now from previous RSU (the vehicle now asks for downloading the remaining data from the current RSU) and deadline is the critical time constraint of the request, beyond which the vehicle move out from RSU area. Each vehicle is equipped with a GPS (Global position System), therefore vehicles know their own geographic position and driving velocity. Then vehicle can estimate it's leaving time, which indeed is the service deadline, mentioned above.

We use the following metric for evaluating the scheduling algorithm:

Service Ratio: Service ratio is defined as the ratio of the number of vehicles which downloaded data in a time limit completely to the total number of vehicle that requested the data. A good scheduling scheme should serve as many requests as possible in time confine.

4   Resuming Download from a RSU

In first step we propose a method for improving service ratio. Each vehicle wants to access to the data from RSU, send a request to the RSU. Requests are queued at the RSU server upon arrival then server serves suitable request base on scheduling scheme. If vehicles move out from a RSU area while downloading data, we have two alternatives:

1. When the vehicle enters the next RSU's coverage area, it request for data again and disregard the amount of data which is already downloaded [11]. This method waste downloaded data and decrease service ratio.

2. Suppose that a vehicle downloaded a part of data from one RSU then it move out from the RSU's coverage area. When the vehicle enters to another RSU's coverage area, it only downloads remaining part of the data.

In the second method server (i.e., the RSU) first should split data into several individual segments and send them to requesting vehicles consecutively. Each vehicle keeps track of the segments it has received. When the vehicle moves out from the RSU and later enters another RSU, it informs the new RSU about the segments which are received. In other word, the vehicle inserts the number of received parts in its request and the RSU just sends remaining segments. It should be noted that all RSUs have the same data (files) and to be synchronized, all RSUs are connected together with a high speed network.

5   The Scheduling Scheme

In this section we assume that downloads can be resumed in each new RSU as discussed in the previous section. We then propose scheduling algorithm based on this assumption.

5.1   Differentiating between Requests in the RSU

In the previous scheduling approach we did not consider any distinction between a vehicle which requests a file from the scratch and one which resumes the download. To take into account this difference, we propose a scheduling scheme based on which requests are lined in two different queues. In case a vehicle requests for a data item for the first time, its request queues in f-queue and if the vehicle requests the remaining part of a data item, its request goes to c-queue. Now the RSU should select a queue taking advantage from a specific scheduling scheme (e.g., FIFO, FCFS, SDF, D*S [11]). It should be stressed that one can add several queues to each RSU for their desired purpose. For instance: a queue for critical requests such as asking ambulance or police, and, another queue for safety requests such as accident notification. Adding several queues to RSU gives the RSU's manager to apply different policies in each queue. The policy for setting queues' priority depends on some parameter such as workload on each queue or queue application and criticality. We suppose that request items in the two queues (f-queue, c-queue) sorted with FCFS scheme, its means that the request with the earliest arrival time in each queue will be served first. In following, we propose methods for selecting a request from head of these queues.

5.2 Scheduling Policies

The primary goal of scheduling scheme is serving as many requests as possible (i.e., increasing the service ratio metric as defined before). We introduce two fields from the 5-cupled request that can be used for the aim of scheduling:

- *Data Size*: if the vehicles can communicate with the RSU at the same data transmission rate, the data size can be taken to decide how long the service will last.
- *Deadline*: Since the request with an earlier deadline is more urgent than the request with a later deadline.

Using the above parameters, our scheduling algorithm includes two following steps as depicted in Figure 2:

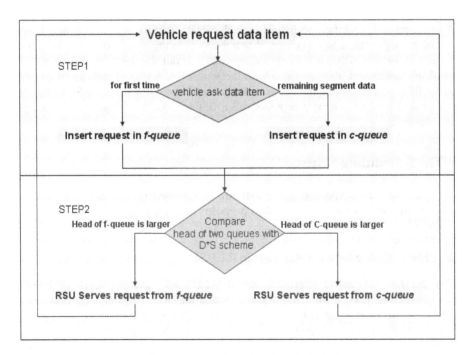

Fig. 2. Flow Chart of the Two-Step Scheduling

1. In first step: requests are separated based on whether the requested data item is asked by the vehicle for the first time or it is a part of resumed request. Requests are inserted in two queues and within each queue they are sorted with FCFS scheduling scheme.

2. In second step: we select one of two queues for the aim of transmission with D*S scheduling scheme.

Note that one may consider different scheduling policies as below:
- First Come First Serve (FCFS): the request with the earliest arrival time will be served first.
- First Deadline First (FDF): the request with the most urgency will be served first.
- Smallest Data Size First (SDF): the data with a small size will be served first.

FCFS does not consider data size and request deadline. FDF does not consider data size thus it ignore the service time. SDF ignores the request urgency. As result, none of them can provide a comprehensive scheduling. With composition of data size and deadline in one scheme an improved scheduling performance is achieved. An instance of the mentioned trend is proposed in [11] under the title of D*S. We select one of the two queues with this method. D*S scheduling rules are as follows:

- Two request items on head of two queues with the same deadline, the one asking for small size data should be served first.
- Two request items on head of two queues with the same size, the one with earlier deadline should be served first.

These parameters compose with following formula [11]:

$$DS\_value = (\ Deadline - CurrentClock\ ) * DataSize \tag{1}$$

In other words, each queue's head which have less DS_value is served first.

6 Performance Evaluations

6.1 Experimental Setup

We simulate our proposed algorithm with ns-2 [13]. The simulation area is a 400*400 square that includes 9 cross-roads. These cross-roads are connected together with horizontal or vertical roads (100 m length) where each two way road has two lanes. We simulate a typical traffic with 130 vehicles in these roads. These vehicles moved with various velocities from 0 to 40 km/h. If vehicle want stop behind the traffic light decrease its velocity and don't stop suddenly.

We can fit the density of vehicle in each cross-road with Poisson distribution function. If we suppose X is a random variable then we can said that X have Poisson distribution with μ parameter that $\mu = \overline{X}$ and $\overline{X}$ is average of number of vehicles in the cross-road. Then probability density function of X is [14]:

$$f(X = x) = \frac{\mu^X * e^{-\mu}}{x!}\quad , x = 0, 1, 2, \dots \tag{2}$$

We compute average of x in each cross-road and then get average of all cross-roads that, it is nearly equal 12. Then

$$f(X = x) = \frac{12^x * e^{-12}}{x!}, \; x = 0, 1, 2\ldots \tag{3}$$

The above equation show how many vehicles presumably exist in the cross-road. Based on the above equation and according to parameter setting in this simulation, we can say that probability of existing 12 vehicles is the highest.

When a vehicle arrives at the cross-road, it can select its way among three different directions with different probabilities or it may stop behind the traffic light. Vehicle travel around the simulation area in different directions and they do not move out form the area. Each vehicle issues one or two request depending on the scenario (as shown see Figure 3). However, for other different service workload scenarios, we can change number of service request. Some other simulation parameters are listed in table1.

Table 1. Simulation Setup Parameters

Parameter	Value
Simulation time	100 sec
Transmission rate	64kb/s
Vehicle velocity	0-40(km/h)
Wireless coverage	100 m
Packet size	1000 Byte
Data item size	Various500,1000,2000(KB)
Routing protocol	DSR
Ratio propagation model	Two ray ground
Antenna model	Omni antenna
Mac type	IEEE 802.11
Traffic type	CBR(UDP)

6.2 Service Ratio Evaluation

Goal of scheduling scheme is to serve as many requests as possible. It is indeed throughput of data distribution system. Figure 3 shows the service ratio of different scheduling algorithms. The X-axis is the simulation time. In this figure three scheduling policies are compared as follows: 1) re-starting downloads form the scratch, 2) restarting downloads from the remaining (one queue) and 3) restarting downloads from the remaining (two queue with D*S scheduling policy). As it follows from the figure, the third policy leads to serving more requests in comparison to the other two schemes. None of vehicles move out from simulation area and they keep cruising until their requested service is completed. Nonetheless, as depicted in Figure 3, the policy based on which vehicles re-start downloading from the scratch is unsuccessful to finish all downloads in the simulation time, therefore its service ratio is lower than two other Algorithms in end of simulation time.

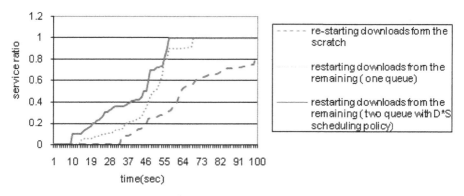

Fig. 3. The service ratio of different scheduling algorithms

6.3 The Effect of Workload

Figure 4 shows the effect of number of request on the scheduling performance for three schemes discussed in this paper. As shown in this figure, when the number of request is increased, the service ratio of third policy descends with lower slope in comparison to the other two scheduling schemes. It follows from the figure that the first policy shows very poor performance. Therefore, the throughput is not scalable if the number of vehicles willing to use data distribution service increases. This actually happens when this service becomes more popular.

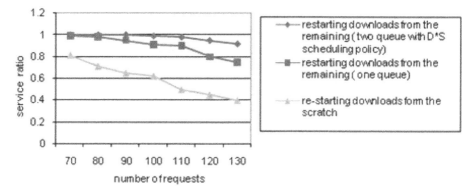

Fig. 4. The effect of workload on service ratio

7   Conclusions

In this paper we proposed scheduling policies for data distribution from RSUs to vehicles. The scheduling policy should be scalable as the number of vehicles as well as the number of files increases. In our proposed approach vehicle are allowed to resume download from a new RSU. In each RSU there are two queues which serve the first-time requests and incomplete requests. Then we select one of two queue with D*S scheduling schemes that consider both service deadline and data size. The results show using D*S scheduling policy along with capability of resumed downloads offer desirable performance and scalability.

References

1. Ott, J., Kutscher, D.: Drive-thru internet: IEEE 802.11b for automobile users. In: Proceedings of Infocom 2004 (2004)
2. Bychkovsky, V., Hull, B., et al.: A measurement study of vehicular internet access using in situ wi-fi networks. In: Proceedings of the 12th annual international conference on Mobile computing and networking (MobiCom 2006) (2006)
3. Hull, B., Bychkovsky, V., Zhang, Y., et al.: Cartel: a distributed mobile sensor computing system. In: Proceedings of the 4th international conference on Embedded networked sensor systems (SenSys 2006), pp. 125–138 (2006)
4. Hadaller, D., Keshav, S., brecht, T., et al.: Vehicular opportunistic communication under the microscope. In: Proceedings of The 5th International Conference on Mobile Systems, Applications, and Services, MobiSys 2007 (2007)
5. Yousefi, S., Fathy, M.: Metrics for performance evaluation of safety applications in vehicular ad hoc networks. Transport. Vilnius: Technika 23(4), 291–298 (2008)
6. Su, C., Tassiulas, L.: Broadcast scheduling for information distribution. In: Proceeding of Infocom 1997 (1997)
7. Aksoy, D., Franklin, M.: R*W: A scheduling approach for large-scale on-demand data broadcast. IEEE/ACM Transactions on Networking 7 (1999)
8. Gandhi, R., Khuller, S., Kim, Y., Wan, Y.: Algorithms for minimizing response time in broadcast scheduling. Algorithmica 38(4), 597–608 (2004)
9. Acharya, S., Muthukrishnan, S.: Scheduling on-demand broadcasts: New metrics and algorithms. In: Proceeding of MobiCom 1998 (1998)
10. Wu, Y., Cao, G.: Stretch-optimal scheduling for on-demand data broadcasts. In: Proceeding of the 10th International Conference on Computer Communications and Networks (ICCCN 2001), pp. 500–504 (2001)
11. Zhang, Y., Zhao, J., Cao, G.: On Scheduling Vehicle-Roadside Data Access. ACM, New York (2007)
12. http://www.csie.ncku.edu.tw/~klan/move/index.htm
13. http://www.isi.edu/nsnam/ns/
14. Schuhl, L., Gerlough, A.: Poisson and Traffic: Use of Poisson Distribution in Highway Traffic & The Probability Theory Applied to Distribution of Vehicles on Two-Lane Highways Daniel

Fast Algorithm of Single Scale Correlative Processing and Its Experimental Research

Ming Xing, Yuan Bingcheng, and Liu Jianguo

Dept. of Weaponry Eng., Naval Univ. of Engineering, Wuhan 430033, China
hgmx503@163.com, yuanbingcheng19502003@yahoo.com,
jianguo_liu8@hotmail.com

Abstract. According to the characteristic of underwater high speed moving target, a signal detection and parameter estimation fast algorithm based on wideband Pro and Con frequency modulation HFM signal was advanced and a simulation on the algorithm was carried on. Through the lake test which simulates the echo of underwater high speed moving target, a DSP system with a core of TMS320C6701 was designed, realizing the fast algorithm of echo processing under the low SNR. The result of lake test indicates that the fast algorithm based on wideband correlative processing and its DSP system can effectively accomplish the detection and parameter estimation on the echo of high speed moving target under low SNR.

Keywords: HFM Signal TMS320C6701 Fast Algorithm Lake Test.

1 Introduction

Detection and parameters estimation on underwater high-speed moving target are greatly focused by researchers. Wideband signal processing is regarded as a proper method for it, but it cause great operation [1]. Adopting HFM signal whose style and processing method are particularly suited to small high speed target, can realize the fast detection and parameter estimation on target, thus a wide attention is aroused on the signal [2],[3].

2 Wideband Correlative Processing

Set the transmitted signal as $f(t)$, received signal as $r(t)$, then

$$r(t) = g(t) + n(t) \tag{1}$$

Where $g(t)$ is echo of target; $n(t)$ is WGN. Suppose that the target keeps an invariable speed and the speed is v, then the model of target echo can be expressed as follows:

$$g(t) = \sqrt{s_0} \cdot f[s_0(t - \tau_0)] \tag{2}$$

Y. Wu and Q. Luo (Eds.): ICHCC-ICTMF 2009, CCIS 66, pp. 31–37, 2010.

s_0 is time stretching according to the target speed, $s_0 = (c-v)/(c+v)$. τ_0 is time delay according to target distance. Wideband correlative processing is the most direct method to signal detection and parameter estimation. The expression of statistic value of detection is shown as follows [1]:

$$\eta = \max_{s_m,\tau_n} \left| \int_0^T r(t) \cdot \sqrt{s_m} \cdot f^*[s_m(t-\tau_n)]dt \right| \mathop{\gtrless}_{<}^{>} \gamma' \qquad (3)$$

s_m and τ_n are discrete time stretching and time delay respectively, γ' is the target detection threshold. s_m and τ_n should be discrete enough to get the max value precisely. But in the meanwhile, this may cause problems such as mass or complex calculating operation. Thus a fast signal processing algorithm is imperatively needed.

3   Fast Algorithm of Single Scale Correlative Processing

3.1   Analysis on Pro and Con frequency Modulation HFM Signal

The expression of HFM signal:

$$f(t) = A(t) \cdot \exp\{ j[2\pi K \ln(1-t/t_0)]\} \qquad (4)$$

Where $A(t)$ is profile function of signal; $K = Tf_i(-T/2)f_i(T/2)/B$, $t_0 = f_m T/B$, B is bandwidth of the signal, T is time length, f_m is the arithmetic center frequency. Considering the Con frequency modulation HFM signal is $f^-(t)$, the Pro frequency signal is $f^+(t)$, and $f^-(t) = f(-t)$, hereby getting the expression of Con frequency modulation HFM signal:

$$f^-(t) = A(t) \cdot \exp\{ j[2\pi K \ln(1+t/t_0)]\} \qquad (5)$$

The Pro and Con frequency modulation HFM signal is shown as follows:

$$f^\pm(t) = f^+(t) + f^-(t) \qquad (6)$$

Where symbol "$\pm$" stands for the overlapping of Pro and Con frequency modulation signal parts. The self-wavelet transformation of Pro and Con frequency modulation HFM signal $f^\pm(t)$ can be gained by making the self-wavelet transformation result of $f^+(t)$ and $f^-(t)$ add together. That is:

$$W_{f^\pm} f^\pm(a,b) = W_{f^+} f^+(a,b) + W_{f^-} f^-(a,b) \qquad (7)$$

$W_{f^\pm} f^\pm(a,b)$ is formed by the peak values of two intersected self-wavelet transformation and shown in Fig.1.

The self-wavelet transformation of the Pro and Con frequency modulation HFM signal is formed by two intersected wave peak values and a sharp peak is formed in center. The signal effectively eliminates the coupling between the time stretching and time delay of unilateral frequency modulation HFM signal and its center peak has an excellent resolution for time stretching and time delay.

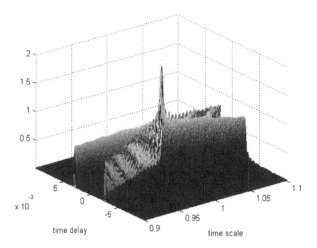

Fig. 1. Self-wavelet transformation of Pro and Con frequency modulation HFM signal

3.2 Fast Algorithm of Single Scale Correlative Processing

The peak values track of self-wavelet transformation of Pro and Con frequency modulation HFM signal is the two intersected beelines on the time stretching and time delay domains:

$$a = 1/s = 1 - b/t_0 \tag{8}$$

$$a = 1/s = 1 + b/t_0 \tag{9}$$

As to the target echo, let $a_0 = 1/s_0$, then the wavelet transformation peak track of transmitted signal is:

$$a = a_0 - (\tau - \tau_0)/t_0 \tag{10}$$

$$a = a_0 + (\tau - \tau_0)/t_0 \tag{11}$$

In perfect situation, the correlative peak track of target echo is two intersected beelines and the position of the point of intersection indicates time delay and time scale of target. The slope of peak track is just related to the transmitted signal for $t_0 = f_m T / B$ and has no relation with the echo's speed and time delay. To take advantage of the characteristic, a fast algorithm of single scale correlative processing is designed. Considering correlative ectypal scale $s = s^*$, expression (10) and (11) can be transformed as:

$$\tau_1 = -(1/s_0 - 1/s^*) \cdot t_0 + \tau_0 \tag{12}$$

$$\tau_2 = (1/s_0 - 1/s^*) \cdot t_0 + \tau_0 \tag{13}$$

Shown in Fig.2, the position of intersection locates at (τ_0, a_0), involving the information of target's time delay and speed.

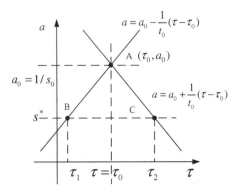

Fig. 2. Sketch map of time delay and scale estimation of single scale correlative processing

As illustrated in Fig.2, there are two intersections B and C, combining with A which is formed by two peak track line intersected, that generates the ΔABC. The acme A reflects the time delay and time stretching of target echo. Thus, according to the position of acme $B(\tau_1, 1/s^*)$, $C(\tau_2, 1/s^*)$, the position of acme A which is the parameter of echo's time delay and time stretching can be easily figured out.

$$\tau_0 = (\tau_1 + \tau_2)/2 \tag{14}$$

$$s_0 = [1/s^* + (\tau_2 - \tau_1)/2t_0]^{-1} \tag{15}$$

The essential of the method is to transform the problem of parameters estimation into a calculation on position of a triangle's acme. The estimation of time delay and time stretching can be accomplished easily with creating one ectype. The operation capacity of method is cut down greatly. So it is appropriate algorithm which can be used in the area of fast detection and parameters estimation.

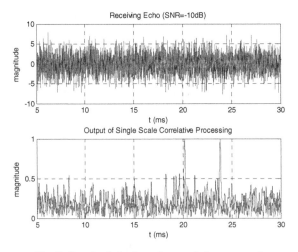

Fig. 3. Result of single scale correlative processing

3.3 The Fast Algorithm Simulation

Assuming the target speed is $v = 80kn = 41.2m/s$, time stretching of the echo is $s_0 = 0.9465$, time delay is $\tau_0 = 0.02s$, noise is WGN, and SNR=-10dB. The correlative processing result is shown in Fig.3:

The signal processing result under SNR=-10dB is illustrated in Fig.3. In this situation, the value of parameter estimation is $\hat{\tau} = 0.0209s, \hat{s} = 0.9477$. The result of simulation indicates that the algorithm is competent for target echo detection and precise parameters estimation under low SNR.

4   Design of DSP System

The core processing chip of the hardware is floating point DSP-TMS320C6701 [4]. The program includes: initialization subprogram, data transfer, correlative processing program, program of parameters estimation and target detection. The flow chart of DSP system is shown in Fig.4:

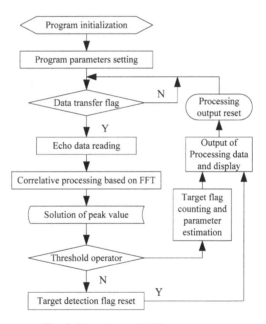

Fig. 4. Flow chart of DSP program

When running the program of system initialization, it should set the parameters for system states control register, external interface of memorizer, and control register and etc. to assure the system in a normal work states. After finishing the initialization, reading data and running correlative processing subprogram runs in turn. Finally, the program of target detection and parameters estimation are finished [4].

5 Lake Test

Due to the difficulty to collect the reflecting echo of underwater high-speed moving target, the test adopts a method which simulates the echo of high speed target to accomplish the dynamic target experiment to testify the effective of algorithm and performance of DSP system. The lake test system is shown in Fig.5:

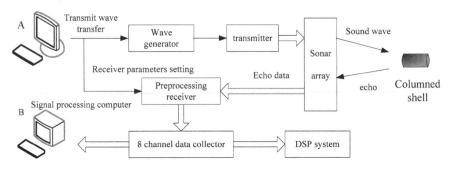

Fig. 5. Lake test signal flow chart

In Lake test, wave generator and transmitter load signal into sonic array. After reflected by columned shell, simulated target echo is received by sonic array, then transferring into preprocessing receiver, inputting into the computer and DSP system to be processed after being collected by mass memorizer.

The distance between columned shell and sonic array is about 141 meters. Consider SNR=-10dB, adopting single scale correlative processing algorithm, the relative speed between simulated target and platform is 80kn. The result of single scale correlative processing algorithm is shown in Fig.6:

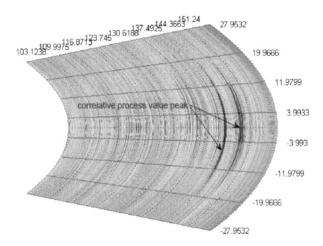

Fig. 6. Result of single scale correlative processing

The position of two bright arcs is the results of the fast algorithm [5],[6]. The estimated target distance value is 141.6 meters, and the estimated velocity value is 76.497kn. Lake test results indicate: under the low SNR, single scale correlative processing algorithm is competent for signal detection and parameter estimation effectively.

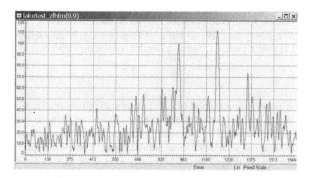

Fig. 7. Processing result of DSP system

Adopting the signal processing method above, we can realize the echo's signal detection and parameter estimation fast algorithm on DSP system [7]. Fig.7 shows the processing result of DSP system.

6   Conclusion

The paper indicates a fast algorithm based on Pro and Con frequency modulation HFM signal. A lake test is designed, simulating the reflecting echo of high-speed target under different SNR, and detection and parameters estimation on target echo is completed based on TMS320C6701 experiment system. The result of simulation and lake test indicates that the fast algorithm is competent for the high-speed target echo signal processing under a low SNR; DSP runs steadily and effectively accomplishes the echo detection and parameter estimation, achieving the anticipated results.

References

1. Van Trees, H.L.: Detection, Estimation, and Modulation Theory. Part I, Part III. Wiley, Chichester (1968/1971)
2. Stein, S.: Differential Delay/Doppler ML Estimation with Unknown Signals. IEEE Trans. Signal Processing 41(8), 2717–2719 (1993)
3. Jin, Q., Wong, K.M., Luo, Z.Q.: The Estimation of Time Delay and Doppler Stretch of Wideband Signals. IEEE Trans. on SP 43(4), 904–916 (1995)
4. Texas Instrument Inc. TMS320C6201/6701 Evaluation Module Technical Reference 1998 564~657 (1998)
5. Frisch, M., Messer, H.: Detection of a Known Tansient Signal of Unknown Scaling and Arrival Time. IEEE Trans. Signal Processing 42(7), 1859–1863 (1994)
6. Kay, S.M.: Fundamentals of Statistical Signal Processing, vol. II. Prentice Hall PTR, Englewood Cliffs (1993)
7. TMS320C6000 CPU and Instruction Set Reference Guide.TI, pp. 265–312 (2000)

A Design Method of Constant Beamwidth Beam-Former

Liu Jianguo[1,2] and Yuan Bingcheng[2]

[1] School of Marine Eng., Northwestern Polytechnical Univ., Xi'an 710072, China
[2] Dept. of Weaponry Eng., Naval Univ. of Engineering, Wuhan 430033, China
Jianguo_liu8@hotmail.com, yuanbingcheng19502003@yahoo.om

Abstract. A new design method of constant beamwidth beam-former (CBB) is proposed: within every frequency sub-band, assume weighting coefficients sequences of beams oriented to different directions are the same, and then it can be concluded that the sequences in all sub-bands have the same spectrum, but a scale relationship is satisfied on the frequency axes. The scale is equal to the ratio of the central frequency in each sub-band to the reference frequency. By use of Chirped Z Transform (CZT), the spectrums of weighting sequences in all sub-bands can be calculated, and then they can be obtained by inverse fast Fourier transform (IFFT). The validity of these conclusions is validated with computer simulations.

Keywords: CZT, Constant Beam-width, Wideband Beam-forming.

1 Introduction

Since more information is obtained with wideband systems, which is more propitious for target detection, parameters estimation and recognition, adopting wideband signals to modern orienting systems has grown into a trend [1]. After filtered within the space field, orthoscopic wideband beam signals can be acquired with CBB. Making use of traditional high-resolution direction of arrival (DOA) estimation methods, new ones in the wideband beam domain are developed. The same as narrowband methods, they have many merits, for example a few operations, high resolution and good allowability property to different tolerance etc.

Many CBB algorithms have been proposed, and their design ideas can be classified into two groups [2]: 1. All sensors are contributing, and weighting coefficients of different frequency are calculated based on some restriction rules. Such methods include Least Square [3], Fourier Transform (FTCBB) [4] and Bessel Function methods [5] etc. 2. Along with the change of frequency, the virtual apertures of the array vary. Linear Combination Array [6] and Spatial Resample (SPARE) [7] etc. are involved. As for most of them, a biased estimate is presented by high resolution methods in the corresponding beam domain. The frequency maximum or minimum of input signal is supposed as the reference frequency in these methods, but an array is not always designed according to the maximum or minimum. With signal bandwidth increment, the number of sub-bands increases, and they will be more complicated to be achieved, and the property of CBB will be also worse.

Y. Wu and Q. Luo (Eds.): ICHCC-ICTMF 2009, CCIS 66, pp. 38–45, 2010.
© Springer-Verlag Berlin Heidelberg 2010

According to the definition of constant beam-width, its mathematical description is given firstly. In every sub-band, weighting coefficients sequences of beams oriented to different direction is supposed to be the same, and then a frequency model of the CBB matrix is represented: the sequences in all frequency sub-bands have the same spectrum, but a scale relationship is satisfied on the frequency axes. The scale is equal to the ratio of the central frequency in each sub-band to the reference frequency. Based on the requirement of wideband beam-forming, a weighting coefficients sequence of the reference frequency is acquired with design methods of narrow-band beam-forming. With the CZT algorithm, spectrums of the sequences in all sub-bands can be fast calculated, and then they can be obtained by IFFT. The validity of these conclusions will be validated by computer simulations.

2 A Frequency Model of the CBB Matrix

Assuming the array is a uniform linear array (ULA) composed of M sensors, the space between adjacent sensors is d. Some pie slice is covered with K beams, and corresponding direction angles are $\alpha_0, \alpha_1, ...\alpha_{K-1}$ respectively. The beam vector of the frequency f oriented to the direction α_k is

$$b(f,\alpha_k) = [b_f(0,\alpha_k),...,b_f(M-1,\alpha_k)e^{j2\pi f(M-1)d\sin\alpha_k/C}]^T \tag{1}$$

where $b_f(0,\alpha_k), b_f(1,\alpha_k),...,b_f(M-1,\alpha_k)$ is called as the corresponding weighting coefficients sequence. The beam-forming matrix is represented as

$$\boldsymbol{B}(f_j) = \left[\boldsymbol{b}(f_j,\alpha_0), \boldsymbol{b}(f_j,\alpha_1),...\boldsymbol{b}(f_j,\alpha_{K-1}) \right], j = 1,2,...J \tag{2}$$

According the CBB definition, the same response in all sub-bands is required

$$\boldsymbol{B}^H(f_j,\boldsymbol{\alpha})\boldsymbol{A}(f_j,\boldsymbol{\theta}) = \boldsymbol{B}^H(f_0,\boldsymbol{\alpha})\boldsymbol{A}(f_0,\boldsymbol{\theta}), j = 1,2,...J \tag{3}$$

where $A(f_j,\boldsymbol{\theta})$ is the array manifold. It is satisfied for $j \in [1,J]$ that

$$\boldsymbol{b}^H(f_j,\alpha_k)\boldsymbol{a}(f_j,\theta_i) = \boldsymbol{b}^H(f_0,\alpha_k)\boldsymbol{a}(f_0,\theta_i) \tag{4}$$

where $k = 0,1,...K-1; i = 0,1,..D-1$. Transform the equation (4) to

$$\sum_{m=0}^{M-1} b_{f_j}^*(m,\alpha_k)e^{-j2\pi f_j x_m(\sin\alpha_k - \sin\theta_i)/C} = \sum_{m=0}^{M-1} b_{f_0}^*(m,\alpha_k)e^{-j2\pi f_0 x_m(\sin\alpha_k - \sin\theta_i)/C} \tag{5}$$

And append new restrictions that every item in the left side is equal to the one in the corresponding position of the right side

$$b_{f_j}^*(m,\alpha_k)e^{-j2\pi f_j x_m(\sin\alpha_k - \sin\theta_i)/C} = b_{f_0}^*(m,\alpha_k)e^{-j2\pi f_0 x_m(\sin\alpha_k - \sin\theta_i)/C} \tag{6}$$

Assume there is a source directly in front of the array, which means $\theta_i = 0$, it is concluded that

$$b_{f_j}(m,\alpha_k) = b_{f_0}(m,\alpha_k)e^{-j2\pi(f_j - f_0)(m-1)d\sin\alpha_k/C} \tag{7}$$

It is called as Frequency Domain Constant Beam-width Beam-former (FDCBB) [8]. However the equation (5) is false when $\theta_i = 0$ or $\theta_i \neq \alpha_k, k \in [0, K-1]$. During the process of high-resolution DOA estimations in the corresponding beam domain, a unbiased estimate can be presented when the DOA is equal to some one among direction angles $[\alpha_0, \alpha_1, \cdots, \alpha_{K-1}]^T$. Since this condition is hardly satisfied in practical applications, only a biased estimate is given by this method.

Since the equation (4) is indeterminate, appending different restrictions, different solutions of weighting coefficients will be obtained. In order to acquire the same beam response oriented to all direction angles in the same frequency sub-band, let $b_{f_j}(m,\alpha_k) = b_{f_j}(m,\alpha_0)$, $m \in [0, M-1]$, then the beam matrix is

$$B(f_j,\alpha) = \Lambda A(f_j,\alpha) \tag{8}$$

where Λ is a diagonal matrix composed by the complex weighting coefficients sequence $[b_{f_j}(0,\alpha_0), b_{f_j}(1,\alpha_0), ..., b_{f_j}(M-1,\alpha_0)]^T$.

Still suppose the special case $\theta_i = 0$, there is

$$\sum_{m=0}^{M-1} b_{f_j}(m,\alpha_0)e^{j2\pi f_j x_m \sin\alpha_k/C} = \sum_{m=0}^{M-1} b_{f_0}(m,\alpha_0)e^{j2\pi f_0 x_m \sin\alpha_k/C} \tag{9}$$

where $k \in [0, K-1]$. Let $f'(\alpha) = f_j \sin\alpha$, it is deduced that

$$\sum_{m=0}^{M-1} b_{f_j}(m,\alpha_0)e^{-j2\pi f'(\alpha_k)md/C} = \sum_{m=0}^{M-1} b_{f_0}(m,\alpha_0)e^{j2\pi f_0'(\alpha_k)md/C} \tag{10}$$

Since α_k can be chosen among $\alpha_0, \alpha_1, ..., \alpha_{K-1}$, the α_k range can be extended to $-90^0 \sim 90^0$. The equation (5) is always right when α_k changes at random in the above range. Keeping the f_j value constant, it can be represented as the DTFT of the discrete sequences $b_{f_j}(m,\alpha_0)$ and $b_{f_0}(m,\alpha_0)$, $m = 0,1,...,M-1$

$$B'(f') = \sum_{m=0}^{M-1} b_{f_j}(m,\alpha_0)e^{-j2\pi f'md/C} = \sum_{m=0}^{M-1} b_{f_0}(m,\alpha_0)e^{j2\pi f_0'md/C} = B(f_0') \tag{11}$$

According to the representation (11), whether the beams oriented to any sources or not, the equation (5) is always satisfied as long as (11) is right. It is meant that a DOA pre-estimate is unnecessary within the CBB methods based on the equation (11). Since

$$f'(\alpha) = f_j \sin\alpha = \left(f_j/f_0 \right) f_0 \sin\alpha = \left(f_j/f_0 \right) f_0'(\alpha) \qquad (12)$$

The equation (11) can be represented as

$$B'(a_j f) = B(f) \quad \text{or} \quad B'(f) = B(s_j f) \qquad (13)$$

where $a_j = f_j/f_0, s_j = f_0/f_j$. It is shown that discrete sequences $b_{f_j}(m,\alpha_0)$ and $b_{f_0}(m,\alpha_0)$ $m = 0,1,...,M-1$ have the same spectrum, while a scale relationship is satisfied on the frequency axes. The scale is equal to the ratio of the central frequency f_j of the j-th sub-band to the reference frequency f_0. Apparently the equation (18) has nothing to do with the reference frequency value.

3 The Design of CBB Based on CZT

It is necessary that more sub-bands are divided in the signal bandwidth range, when the bandwidth is very wide, or its spectrum undulation is intensive. Scales s_j may be fractions or irrational numbers in this case, and it will be very difficult to calculate the weighting coefficients based on FTCBB or SPARE. As for any value of the scale s_j, as long as the sequence $b_{f_0}(m,0), m = 0,1,...,M-1$ of the reference frequency is known, $B(s_j f')$ can be calculated by use of CZT quickly. Then the sequence $b_{f_j}(m,0), m = 0,1,...,M-1$ is obtained with IFFT.

3.1 Chirped Z Transform (CZT)

If $x(n)$ is a known discrete signal, its CZT can be represented as

$$X(z) = \sum_{n=0}^{+\infty} x(n) z^{-n} \qquad (14)$$

where $z = e^{sT_s} = e^{\sigma T_s} e^{j\Omega T_s} = Ae^{j\omega}$. Let $z_k = AW^{-k}$, where $A = A_0 e^{j\theta_0}$, $W = W_0 e^{-j\varphi_0}$, there is

$$z_k = A_0 e^{j\theta_0} \left(W_0^{-k} e^{jk\varphi_0} \right) \qquad (15)$$

Pre-establishing the value of parameters $A_0, \varphi_0, W_0, \theta_0$, there is a dot track of z_k, $k = 0,1,...M-1$ in the z plane. It is a segment of arc between the starting point

$z_0 = A_0 e^{j\theta_0}$ with the end-point $z_{M-1} = A_0 e^{j\theta_0} W_0^{M-1} e^{j(M-1)\phi_0}$. These M points are located evenly on the arc. The CZT value at these points can be calculated as

$$X(z_k) = \text{CZT}(x(n)) = \sum_{n=0}^{M-1} x(n) z_k^{-n} \tag{16}$$

3.2 Calculation of CBB Matrix Based on CZT

Transform the equation (13) to a discrete representation

$$B'(2\pi m/M) = B(2\pi s m/M), m = 0,1,...M-1 \tag{17}$$

It is presented that $B'(f)$ are equal to the values of $B(f)$ at the positions evenly distributed in the range $-s\pi \sim +s\pi$, shown as Fig.1 (where $s > 1$). To describe the track clearly, $A_0 > 1$ is supposed. The values of $B'(2\pi m/M)$, $m \in [0, M-1]$ can be obtained by DFT at some segment of the unit circle. CZT is competent for this task, whose character is to do DFT at any segment of the unit circle nicely.

Firstly it is necessary to choose any value at the signal bandwidth range as the reference frequency f_0 before the CBB weighting matrix is calculated with CZT. Actually it is determined by the space between two adjacent sensors. According to the requirement of wideband beam-forming, for example the width of main lobe, side-lobe level and null steering position, the sequence $b_{f_0}(m,0), m = 0,1,..., M-1$ is given with design methods of narrow beam-forming.

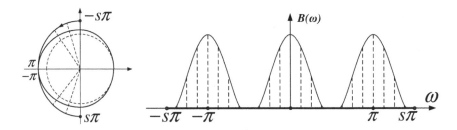

Fig. 1. The sketch map of the design of CBB based on CZT

Since the sensors number is usually small, the weighting coefficients sequence $b_{f_0}(m,0)$, $m = 0,1,..., M-1$ is extended to L points by Zero-Padding in order to decrease the calculation errors, where L is a exponent of 2. Let

$$A = e^{-j\pi s_j}, \quad W = e^{j2\pi s_j/L} \tag{18}$$

$B'(2\pi m/L)$, $m = -L/2 \sim L/2 - 1$ are calculated by CZT algorithm, then the first half is exchanged with the remaining half. The new sequence is transformed to $b'_{f_j}(m,0)$ $m = 0 \sim L-1$ by IFFT. The first M data is truncated, and regarded as the sequence $b_{f_j}(m,0)$, $m = 0 \sim M - 1$. Calculating sequences of all scales, all CBB matrixes $\boldsymbol{B}(f_j, \boldsymbol{a})$, $j \in [1, J]$ are obtained with the equation (8).

It is presented above that the frequency model of CBB weighting matrix has nothing to do with the reference frequency value. Any value within the signal band-width range can be chosen as the reference when this method is used. In order to satisfy the space sampling theorem, the frequency maximum can be regarded as the reference, where all scales are larger than 1. While liking SPARE, the frequency minimum can be also chosen as the reference to make sure that all weighting sequences are inside of the effective aperture of the array. Since a DOA pre-estimate is unnecessary with this method, an unbiased estimate can be acquired by high-resolution DOA estimation methods in the beam domain based on this method.

4 Computer Simulation

Computer simulation results based on the new method are presented in the section. Simulation 1: the bandwidth range is [20, 30] kHz. The minimum is chosen as the reference frequency, which means all scales are less than or equal to 1. The ULA is composed of 32 isotropy sensors, and the sensors space is equal to a half wavelength corresponding to the reference frequency. Even weighting is adopted as the sequence $b_{f_0}(m,0), m = 0,1,...,31$. CBB are designed with the method above and SPARE, and their beam patterns are shown as fig.2. It is given that compared to SPARE, not only good constant bandwidth performance is ensured, but also better consistencies of side-lopes can be acquired with the method given in this paper.

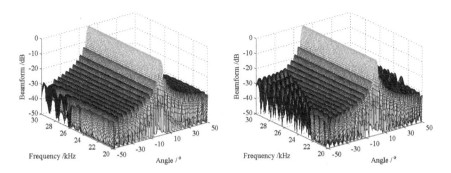

Fig. 2. the beam patterns of CBB. The left one is based on CZT, and the other is SPARE.

Simulation 2: the frequency ranges are [10, 30] and [5 30] kHz respectively. The maximums are chosen as the reference frequencies. The ULA is composed of 16 isotropy sensors, and the space between adjacent sensors is equal to a half wavelength corresponding to the reference frequency too. Chebyshev weighting is adopted as the sequence $b_{f_0}(m,0)$, $m \in [0,31]$. The beam patterns are shown as Fig.3.

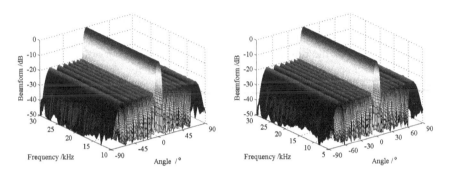

Fig. 3. the beam patterns of CBB with different frequency bandwidths

It is shown that with increment of signal bandwidth, nearly perfect constant bandwidth performance is still obtained, and the all beam-forms are almost consistent.

5   Conclusion

In this paper, a mathematical description of CBB is presented, which is an indeterminate equation. New restrictions are appended that within every frequency sub-band, assume weighting coefficients sequences of beams oriented to different directions are the same. Then it can be concluded that the sequences within different sub-bands have the same spectrum, while a scale relationship is satisfied on the frequency axes. The scale is equal to the ratio of the central frequency of the sub-band to the reference frequency. As long as the weighting sequence of the reference frequency is known, all sequences in other sub-bands can be obtained by use of CZT. With this method, not only good constant beam-width performance is presented, but also better consistencies of the side-lope are shown.

In many documents on wideband CBB, it is supposed that the frequency minimum or maximum of the input signal is chosen as the reference frequency. However any value at the frequency range of the signal can be regarded as the reference frequency with the new method, which increases good agility for the design and application of the array undoubtedly. A pre-estimate is unnecessary for high resolution DOA estimation in the beam domain based on this method, and an unbiased estimate can be acquired. Validity of the theoretical analysis has been proven by computer simulations. Since the fast algorithm CZT is utilized, the operations to realize the above method is reduced greatly, and the load of hardware systems is also decreased.

References

1. Vaccaw, R.J. (ed.): The Past, Present, and Future of Underwater Acoustic Signal Processing. IEEE Signal Processing Mag., 21–51 (July 1998)
2. Guibing, L.: Design theory of sonar array. Ocean Publishing Company, Beijing (1995)
3. Lee, T.S.: Efficient wideband source localization using beamforming invariance technique. IEEE Trans. Signal Processing 42(6), 1376–1387 (1994)
4. Godara, L.C.: Application of the fast Fourier transform to broadband beamforming. J. Acoust. Soc. Am. 98(1), 230–240 (1995)
5. Yixin, Y., Chao, S.: A new method of broadband constant beamwidth beamforming for arbitrary geometry arrays. Acta Acoustica 26(1), 55–58 (2001)
6. Smith, P.R.: Constant beamwidth receiveing arrays for broadband sonar systems. Acoustica 23, 21–26 (1970)
7. Sidorovich, D.V., Gershman, A.B., et al.: Processing of experimental seismic array data using 2-D wideband interpolated Root-MUSIC. In: Proc. IEEE ICASSP, vol. 4, pp. 1985–1988 (1998)
8. Do-Hong, T., Russer, P.: Wideband direction-of-arrival estimation and beamforming for smart antennas system. In: Proc. of IEEE Radio and Wireless Conf., pp. 127–130 (2003)

Network Service Incentive Mechanism Based on Competitive Game

Zhide Chen[1,2], Jinhua Zheng[1,2], and Li Xu[1,2]

[1] School of Mathematics and Computer Science, Fujian Normal University
[2] Key Laboratory of Network Security and Cryptology, Fujian Normal University
Fuzhou City, 350007
{zhidechen,jhzheng,xuli}@fjnu.edu.cn

Abstract. In this paper, we propose the schemes of network service incentive mechanism based on competitive game. It is supposed that both the management center and the server nodes are selfish and rational. Firstly, we compose a simple incentive mechanism with a management center and two server nodes, and calculate the optimal payoff for the management center and the server nodes. Secondly, we extend the mechanism to one management center and n server nodes, and calculate the optimal payoff for the nodes. The incentive mechanism can be applied to P2P services, routing, security, QoS and so on.

Keywords: Competitive Game, Network Service, P2P, Incentive Mechanism.

1 Introduction

1.1 Network Services

There are several kinds of network services, P2P, routing, security and so on [9,8,6]. Any network services need some kinds of incentive. Take P2P service as a example. In the P2P model, all the peer nodes are equal, one node can get the service from other nodes, and it can also provide the services for the other nodes. The peer nodes need the incentive mechanism to motivate the peer nodes to server the other nodes. Only by this way, can the P2P network improve the performance [4].

1.2 Incentive Mechanism

In network service, it is hard for the server nodes to share their resource or provide the network services without any feedback. Without the incentive mechanism, as the server nodes is reasonable and selfish, the optimal strategy is to maximize their income. The selfish nodes will get the resources or services from other nodes, and do not provide the resources or services for the other nodes. Therefore, the incentive strategy is essential to the network services. Many researchers have their working on these areas.

Incentive Mechanism Based on Payment

In the incentive mechanism based on payment, the resources prices or the services prices is taken as a tools to measure the value of the resources or the services [7,1]. The virtual currency is the medium in the resources or services exchange. The service

Y. Wu and Q. Luo (Eds.): ICHCC-ICTMF 2009, CCIS 66, pp. 46–53, 2009.

providers can get the virtual currency by selling resources or services. The nodes have to pay the virtual currency if they get the resources or services from other nodes. By this way, the virtual currency is the feedback to the resource providers or the services providers. The nodes have the enthusiasm to take part in the cooperation.

Incentive Mechanism Based on Credit

In the incentive mechanism based on credit, every node has the credit value for the neighbor nodes. In each services, the nodes can have evaluate the other nodes with positive credit or negative credit. If one node provides the network services, the credit value would be increased. Otherwise, if one node refuses the resources or the services, the credit value would be decrease. The other nodes would judge whether to provide the network services to the node by his credit value. Therefore, in incentive mechanism based on credit, the nodes have to provide the resources or the services, Otherwise, they cannot get the resources or the services from other nodes [11,13].

Incentive Mechanism Based on Inherit Algorithm

In incentive mechanism based on inherit algorithm, the network service providers allocate the resources or the services based on the inherit algorithm. The goal is to maximize the contribution value. The larger of the contribution value, more resources or services the node can get from the sharing resources or services. These would realize the fairness and the efficiency in the mechanism, and restrain the selfishness [12,10].

Incentive Mechanism Based on Game Theory

In incentive mechanism based on game theory, every node is assumed to be rational and selfish. Their goal is to maximize the payoff. In classic game theory, Nash equilibrium implied that the participants cannot get larger payoff by bias the equilibrium strategy. The goal is to can get the individual optimization and the overall optimization. These means the fairness and the efficiency in the mechanism. It would be the solution for the resources or services allocation [5,2].

1.3   Ideal of This Paper

The goal of this paper is to design a mechanism to motivate the service nodes to better sever the other nodes, and provide the services with 100% effort. For example, two server nodes have the have the same ability to provide services for the other nodes. The max ability is 100. The first node server the other nodes with 100% effort. The second node server the other nodes with 10%. If all the nodes in the network choose the strategy of the second nodes, This would reduce the performance of the network. Therefore, the ideal of this paper is to design a incentive mechanism. The goal is to motivate the server nodes to provide the other nodes with 100% effort.

1.4   Organization

In this paper, we propose a incentive mechanism based on competition game theory, discuss the incentive mechanism for the network service. In section 2, we establish the competition game model for network service with one management center and two competitive server nodes. In section 3, we discuss a model of one management center and n competitive server nodes. In section 4, we provide the conclusion and further work.

2   One Management Center and Two Competitive Server Nodes Incentive Mechanism

2.1   Incentive Model

We first compose the model with one management center and two server nodes based on the competitive game theory [3]. Assuming there is a management center in the network service, its duty is to monitor the server nodes to provide the service. In this model, node 1 and node 2 are two competitive server nodes. The output of the services from the two nodes is marked is

$$y_i = e_i + \varepsilon_i$$

e_i is the effort value, and ε_i is the random factor.

We have three reasonable hypothesists for the scheme. Firstly, the effort value is positive, i.e. $e_i \geq 0$. Secondly, the random factor ε_1 and ε_2 are independent, the expectation of the two variables is 0, the density function is $f(\varepsilon)$. Thirdly, all the services provided by the two nodes can be observed, but the effort value cannot be observed. The payoff is decided by the output of their services, and are not only decided by the effort value.

The management center motivate the two server nodes to provide the services. The competition is between the two server nodes. The successful node will obtain the payoff of w_H, the defeat node will obtain the payoff of w_L. The average payoff for the server nodes is w, the effort value is e. The server nodes' utility function is $\mu(w, e) = w - g(e)$. $g(e)$ is the negative factor in the service. It is a increasing convex function ($g'(e) > 0$ and $g''(e) > 0$). The payoff of the management center is

$$y_1 + y_2 - w_H - w_L.$$

2.2   Analysis

The management center choose the return payoff w_H and w_L. The two server nodes have the effort value e_H and e_L. Both nodes have have the random factor ε_H and ε_L. We discuss the exception payoff.

The management center choose the payoff w_H and w_L. If the effort value (e_1^*, e_2^*) is the Nash equilibrium in the next stage. For every i, e_i^* is the maximum of the net payoff. The value is the expectation payoff minus the negative utility. e_i^* should satisfy

$$
\begin{aligned}
&\max_{e_i > 0} w_H P\{y_i(e_i) > y_i(e_j^*)\} + w_L P\{y_i(e_i) \leq y_i(e_j^*)\} - g(e^*) \\
&= \max_{e_i > 0} w_H P\{y_i(e_i) > y_i(e_j^*)\} + w_L (1 - P\{y_i(e_i) > y_i(e_j^*)\}) - g(e^*) \\
&= (w_H - w_L) P\{y_i(e_i) > y_i(e_j^*)\} + w_L - g(e^*)
\end{aligned}
\tag{1}
$$

The first order condition for

$$y_i(e_i) = e_i + \varepsilon_i$$

is

$$(w_H - w_L) \frac{\partial P\{y_i(e_i) > y_i(e_j^*)\}}{\partial e_i} = g'(e_i^*) \tag{2}$$

That is, when node i select the effort value e_i, the extra effort's negative marginal utility is equal to the marginal benefit. The marginal benefit is equal to the success payoff $w_H - w_L$ multiple the probability of the success.

By the Bayes Law conditional probabilities:

$$P(A|B) = \frac{P(A, B)}{P(B)}.$$

We have

$$
\begin{aligned}
P\{y_i(e_i) > y_j(e_j^*)\} &= P\{\varepsilon_i > e_j^* + \varepsilon_j + e_i\} \\
&= \int_{\varepsilon_j} P\{\varepsilon_i > e_j^* + \varepsilon_j - e_i | \varepsilon_j\} f(\varepsilon_j) d\varepsilon_j \\
&= \int_{\varepsilon_j} [1 - F(e_j^* + \varepsilon_j - e_j)] f(\varepsilon_j) d\varepsilon_j
\end{aligned}
\tag{3}
$$

The first order condition of Equation 2 can be transformed to

$$(w_H - w_L) \int_{\varepsilon_j} [1 - F(e_j^* + \varepsilon_j - e_j)] f(\varepsilon_j) d\varepsilon_j = g'(e_i)$$

In the symmetrical Nash equilibrium (i.e. $e_1^* = e_2^* = e^*$), we have

$$(w_H - w_L) \int_{\varepsilon_j} f^2(\varepsilon_j) d\varepsilon_j = g'(e_i^*) \tag{4}$$

On one hand, as $g(e)$ is a convex function, the lager of the reward for the success node, the larger effort can be motivated. On the other hand, with the same level of the reward, the larger of the random factor, the less effort should provide to server the other nodes.

If ε subject to normal distribution with variance of σ^2, we have

$$\int_{\varepsilon_j} f^2(\varepsilon_j) d\varepsilon_j = \frac{1}{2\sigma \sqrt{\pi}}$$

It decrease when σ increase. i.e. e^* decrease when σ increase.

Assuming the two nodes agree to participate in the services providing competition. Their strategy to the w_H and w_L is the symmetrical Nash equilibrium in Equation 4. The probability that the nodes win the competition is 0.5. If the management center can motivate the nodes to take particular in the competition, the choice must satisfy the following requirement,

$$\frac{1}{2}w_H - \frac{1}{2}w_L - g(e^*) \geq U_a \tag{5}$$

If U_a is to let the management center to motivate the nodes to take particular in the competition, he will choose $2e^* - w_H - w_L$ as the largest payoff under the condition Inequation 5. The equality is established under the optimal condition.

$$w_L = 2U_a + 2g(e^*) - w_H \tag{6}$$

The expected profit is $2e^* - 2U_a + 2g(e^*)$. The management center 's goal is to maximize $e^* - g(e^*)$. He can choose the level of the reward to let e^* satisfy the condition.

Under the optimal condition $g'(e^*) = 1$, it implies

$$(w_H - w_L) \int_{\varepsilon_j} f^2(\varepsilon_j)d\varepsilon_j = 1$$

We can solve the value of w_H and w_L,

$$w_H = U_a + g(e^*) + \frac{1}{2\int_{\varepsilon_j} f^2(\varepsilon_j)d\varepsilon_j}$$

$$w_L = U_a + g(e^*) - \frac{1}{2\int_{\varepsilon_j} f^2(\varepsilon_j)d\varepsilon_j}$$

3   One Management Center and n Competitive Server Nodes Incentive Mechanism

3.1   Incentive Model

Assumed that the average payoff for the server nodes is w, the effort value is e. The server nodes' utility function is

$$\mu(w, e) = w - g(e).$$

$g(e)$ is the negative factor in the service.

To motivate the n server nodes to provide the services, there are competition among the nodes. The rewards of the competition is w_i $(i = 1, 2, \cdots, n)$, supposed that

$$w_1 \leq w_2 \leq \cdots \leq w_n.$$

The average for the server nodes is w, the effort value is e, $\mu(w, e) = w - g(e)$ $(g'(e) > 0$ and $g''(e) > 0$). The payoff of the management center is

$$y_1 + y_2 + \cdots + y_n - (w_1 + w_2 + \cdots + w_n)$$

3.2   Analysis

Assuming the management center is the participator 0, his action is to choose the the reward level w_i $(i = 1, 2, \cdots, n)$, the n nodes are the participator $1, 2, \cdots, n$. They have the information of the reward level, select the action $a_1, a_2, \cdots, a_n$, and decide the effort value. The rewards are decided not only by the actions, but also by the random factor $\varepsilon_1, \varepsilon_2, \cdots, \varepsilon_n$.

Assuming the management center select the reward value w_i $(i = 1, 2, \cdots, n)$, and $w_1 \leq w_2 \leq \cdots \leq w_n$, in the first stage, let the effort value to be $(e_1^*, e_2^*, \cdots, e_n^*)$ is the Nash equilibrium in the next stage. For every i, e_i^* would enable the nodes to maximize the net payoff. The value is the expectation payoff minus the negative utility. We realign the $y_i(e_j^*)$ from small to large. The subscript is changed so that

$$y_i(e_{j1}^*) < y_i(e_{j2}^*) < \cdots < y_i(e_{j(n-1)}^*)$$

$jk \in \{1, \cdots, i-1, i+1, \cdots, n\}$, $k = 1, 2, \cdots, n-1$, e_i^* must satisfy

$$\max_{e_i \geq 0}\{w_n P\{y_i(e_i) > y_i(e^*_{j(n-1)})\}$$
$$+w_{n-1}P\{y_i(e^*_{jn-2}) < y_i(e_i) \leq y_i(e^*_{jn-1})\}$$
$$+\cdots$$
$$+w_2 P\{y_i(e^*_{j1}) < y_i(e_i) \leq y_i(e^*_{j2})\}$$
$$+w_1 P\{y_i(e^*_i) < y_i(e^*_{j1})\}$$
$$-g(e_i)\}$$
$$= \max_{e_i \geq 0}\{w_n P\{y_i(e_i) > y_i(e^*_{j(n-1)})\}$$
$$+w_{n-1}\{P\{y_i(y_i(e_i) < y_i(e^*_{jn-1})\} - P\{y_i(y_i(e_i) \leq y_i(e^*_{jn-2})\}\}$$
$$+\cdots$$
$$+w_2\{P\{y_i(e_i) < y_i(e^*_{j2})\} - P\{y_i(e_i) \leq y_i(e^*_{j1})\}\}$$
$$+w_1 P\{y_i(e^*_i) < y_i(e^*_{j1})\}$$
$$-g(e_i)\} \tag{7}$$
$$= \max_{e_i \geq 0}\{w_n P\{y_i(e_i) > y_i(e^*_{j(n-1)})\}$$
$$+w_{n-1}\{1 - P\{y_i(y_i(e_i) > y_i(e^*_{jn-1})\} - 1 + P\{y_i(y_i(e_i) > y_i(e^*_{jn-2})\}\}$$
$$+\cdots$$
$$+w_2\{P\{1 - y_i(e_i) > y_i(e^*_{j2})\} - 1 + P\{y_i(e_i) > Xy_i(e^*_{j1})\}\}$$
$$+w_1 P\{y_i(e^*_i) < y_i(e^*_{j1})\}$$
$$-g(e_i)\}$$
$$= \max_{e_i \geq 0}\{(w_n - w_{n-1})P\{y_i(e_i) > y_i(e^*_{j(n-1)})\}$$
$$+(w_{n-1} - w_{n-2})P\{y_i(e_i) > y_i(e^*_{j(n-2)})\}$$
$$+\cdots$$
$$+(w_2 - w_1)P\{y_i(e_i) > y_i(e^*_{j1})\}$$
$$+w_1 - g(e_i)\}$$

The first order condition for $y_i(e_i) = e_i + \varepsilon_i$, $i = 1, 2, \cdots, n$ is

$$(w_n - w_{n-1})\frac{\partial P\{y_i(e_i) > y_i(e^*_{j(n-1)})\}}{\partial e_i}$$
$$+ \cdots$$
$$+ (w_2 - w_1)\frac{\partial P\{y_i(e_i) > y_i(e^*_{j1})\}}{\partial e_i} \tag{8}$$
$$= g'(e_i)$$

That is, if node i select the effort value e_i, the extra effort's negative marginal utility is equal to the marginal benefit. The marginal benefit is equal to the success payoff $w_k - w_{k-1}$ $(k = 2, \cdots, n)$ multiple the probability of the success.

By the Bayes Law conditional probabilities, we have

$$P\{y_i(e_i) > y_i(e^*_j)\}$$
$$= \int_{\varepsilon_j} P\{\varepsilon_i > e^*_j + \varepsilon_j - e_i | \varepsilon_j\} f(\varepsilon_j) d\varepsilon_j \tag{9}$$
$$= \int_{\varepsilon_j} [1 - F(e^*_j + \varepsilon_j - e_j)] f(\varepsilon_j) d\varepsilon_j$$

The first order condition of Equation 8 can be transformed to

$$(w_n - w_{n-1}) \int_{\varepsilon_{j(n-1)}} f(e^*_{j(n-1)} - e_i + \varepsilon_{j(n-1)}) f(\varepsilon_{j(n-1)}) d\varepsilon_j$$
$$= \cdots$$
$$= (w_2 - w_1) \int_{\varepsilon_{j1}} f(e^*_{j1} - e_i + \varepsilon_{j1}) f(\varepsilon_{j1}) d\varepsilon_{j1} \tag{10}$$
$$= g'(e_i)$$

In the symmetrical Nash equilibrium (i.e. $e_1^* = e_2^* = \cdots = e_n^* = e^*$), we have

$$
\begin{aligned}
&(w_n - w_{n-1}) \int_{\varepsilon_{j(n-1)}} f^2(\varepsilon_{j(n-1)}) d\varepsilon_j \\
&= \cdots \\
&= (w_2 - w_1) \int_{\varepsilon_{j1}} f^2(\varepsilon_{j1}) d\varepsilon_{j1} \\
&= g'(e^*)
\end{aligned}
\tag{11}
$$

On one hand, as $g(e)$ is a convex function, the lager of the reward for the success node, the larger effort can be motivated. On the other hand, with the same level of the reward, the larger of the random factor, the less effort should provide to server the other nodes.

Let the w_i ($i = 1, 2, \cdots, n$) is a arithmetic progression with common difference d. The reward level w_i ($i = 1, 2, \cdots, n$). Their strategy is the symmetrical Nash equilibrium in Equation 11. The probability that the nodes win the competition is $\frac{1}{n}$ (i.e. $P\{y_i(e_i) > y_i(e^*)\} = \frac{1}{n}, j = 1, \cdots, i-1, i+1, \cdots, n$). If the management center can motivate the nodes to take particular in the competition, the choice must satisfy the following requirement.

$$
\frac{1}{n}(w_1 + w_2 + \cdots + w_n) - g(e^*) \geq U_a
\tag{12}
$$

If U_a is to let the management center to motivate the nodes to take particular in the competition, he will choose $ne^* - (w_1 + w_2 + \cdots + w_n)$ as the largest payoff under the condition Inequation 12. The equality is established under the optimal condition.

$$
(w_1 + w_2 + \cdots + w_n) = 2U_a + 2g(e^*)
$$

The expected profit is $ne^* - 2U_a + 2g(e^*)$. The management center's goal is to maximize $e^* - g(e^*)$. He can choose the level of the reward to let e^* satisfy the condition. Under the condition $g'(e^*) = 1$ of optimal choice, this implies the optimal

$$
\begin{aligned}
&(w_n - w_{n-1}) \int_{\varepsilon_{j(n-1)}} f^2(\varepsilon_{j(n-1)}) d\varepsilon_j \\
&= \cdots \\
&= (w_2 - w_1) \int_{\varepsilon_{j1}} f^2(\varepsilon_{j1}) d\varepsilon_{j1} \\
&= g'(e^*)
\end{aligned}
\tag{13}
$$

That is

$$
\begin{aligned}
&d \int_{\varepsilon_{j(n-1)}} f^2(\varepsilon_{j(n-1)}) d\varepsilon_j \\
&= \cdots \\
&= d \int_{\varepsilon_{j1}} f^2(\varepsilon_{j1}) d\varepsilon_{j1} \\
&= g'(e^*)
\end{aligned}
\tag{14}
$$

We can solve the value of w_i ($i = 1, 2, \cdots, n$).

4   Conclusion and Further Work

In this paper, firstly, we consult the competition game theory, and establish the incentive model of one management center and two nodes. Secondly, we extend this scheme to one management center and n nodes. It provide another solution to incentive for network service. Every nodes in the scheme have the contribution for the network, the reward is optimization.

In the incentive mechanism, every nodes is glad to participate in the game, and reach the goal of entire optimization and individual optimization. In the one management center and n nodes, we hypothesis that the reward level is arithmetic progression with common difference d. In some particular occasion, the difference of the reward level is rational and necessary. These problems is worth of further research.

References

1. Dai, X., Grundy, J.C.: Off-line micro-payment system for content sharing in P2P networks. In: Chakraborty, G. (ed.) ICDCIT 2005. LNCS, vol. 3816, pp. 297–307. Springer, Heidelberg (2005)
2. Ferretti, S.: Cheating detection through game time modeling: A better way to avoid time cheats in P2P MOGs? Multimedia Tools Appl. (MTA) 37(3), 339–363 (2008)
3. Gibbons, R.: A Primer in Game Theory. Harvester Wheatsheaf, London (2001)
4. Horne, B.G., Pinkas, B., Sander, T.: Escrow services and incentives in peer-topeer networks. In: ACM Conference on Electronic Commerce 2001, pp. 85–94 (2001)
5. Lagesse, B., Kumar, M.: A Novel Utility and Game-Theoretic Based Security Mechanism for Mobile P2P Systems. In: PerCom 2008, pp. 486–491 (2008)
6. Lin, M.-H., Lo, C.-C.: A Quality of Relay-Based Incentive Pricing Scheme for Relaying Services in Multi-hop Cellular Networks. In: Kim, C. (ed.) ICOIN 2005. LNCS, vol. 3391, pp. 796–805. Springer, Heidelberg (2005)
7. Liu, Y., Zhao, Z.: Payment Scheme for Multi-Party Cascading P2P Exchange. In: I3E 2007, pp. 560–567 (2007)
8. Ma, R.T.B., Lee, S.C.M., Lui, J.C.S., Yau, D.K.Y.: Incentive and service di_erentiation in P2P networks: a game theoretic approach. IEEE/ACM Trans. Netw (TON) 14(5), 978–991 (2006)
9. Podlesny, M., Gorinsky, S.: Rd network services: differentiation through performance incentives. In: SIGCOMM 2008, pp. 255–266 (2008)
10. Tan, K.C., Wang, M.L., Peng, W.: A P2P genetic algorithm environment for the internet. Commun. ACM (CACM) 48(4), 113–116 (2005)
11. Tian, H., Zou, S., Wang, W., Cheng, S.: A Group Based Reputation System for P2P Networks. In: Yang, L.T., Jin, H., Ma, J., Ungerer, T. (eds.) ATC 2006. LNCS, vol. 4158, pp. 342–351. Springer, Heidelberg (2006)
12. Wong, W.Y., Lau, T.P., King, I.: Information retrieval in P2P networks using genetic algorithm. In: WWW (Special interest tracks and posters) 2005, pp. 922–923 (2005)
13. Wang, W., Zeng, G., Yuan, L.: A Semantic Reputation Mechanism in P2P Semantic Web. In: Mizoguchi, R., Shi, Z.-Z., Giunchiglia, F. (eds.) ASWC 2006. LNCS, vol. 4185, pp. 682–688. Springer, Heidelberg (2006)

An Efficient Computation Saving Mechanism for High-Performance Large-Scale Watershed Simulation

Cheng Chang, Wei-Hong Wang, and Rui Liu

HP Labs China
{chang.cheng,weihong.wang,rui.liu}@hp.com

Abstract. Computer simulation is one of the most powerful tools that scientists have at their disposal for studying sophisticated processes or phenomena, which are infeasible, too costly, or dangerous to reproduce. As simulation programs are often built on complex mathematical models, it is extremely desirable to greatly reduce simulation time at the minimum loss of accuracy, especially when a long simulation time span is dictated or immediate response is demanded. This paper presents our research work on improving computational efficiency of a data-intensive parallel hydraulic analysis application, which targets real-time fine-grained simulation for large-scale watersheds. To deliver computational results at low latency and high throughput, we leverage the processing capabilities of both database and compute cluster, and have designed an interpolation mechanism with which previous results are efficiently managed and dynamically maintained for maximum reuse. Experiments have demonstrated the benefits of our solution.

Keywords: Scientific Computation, Optimization, Function Approximation, Database, Simulation.

1 Introduction

Simulation is one of the most powerful tools that scientists have at their disposal for studying complex processes or phenomena. Simulation programs are coded from mathematical models that are established according to relevant physical laws. As scientists integrate more sophisticated models into simulations, computational cost of these computerized mathematical models can be increasingly standing out. For example, in a typical hydraulic simulation task we have investigated, a detailed river model for an area of a few dozens of square kilometers could involve thousands of river segments, each incurring a number of nonlinear partial differential equations, and connected segments are strongly correlated spatially and temporally.

We address the key issue to greatly enhance throughput and latency for complex scientific simulations that run for many simulation time steps, with large-scale watershed simulation as an example. Our work constitutes part of a research project called Hydro-Earth. The project aims to offer fine-grained hydraulic information services, for both public and specialized users, which can be reduced to database queries for hydraulic metrics that either extend over a time span or correspond to some real-time inputs. The metrics are simulated results of yielding and diffusion processes of water and sand within all river segments of a watershed.

Y. Wu and Q. Luo (Eds.): ICHCC-ICTMF 2009, CCIS 66, pp. 54–61, 2010.

To deliver high volumes of results with low latency, we need to resolve two major challenges: (1) spatial and temporal dependency, and (2) repetitive computation. For a river, yielding and diffusion of water and sand at the downstream side are always dependent on those at the upstream side. Thus, computation on downstream river segments cannot be started until that on relevant upstream segments is already finished, raising issues of delayed response and long execution time. Regardless of the time span of the simulation request (a time range or a specific time point), simulation must be performed over a significant simulation time span to build up an accountable history of the simulated system up to the time point requested, and to simulate over the specified range as well. As a result, the inherently nonlinear and intertwined equations are solved repetitively for a great many times, although under different combinations of coefficients, which reemphasizes the issue of computational cost.

In this paper, we focus on exploring strategies for significant reduction of repeated computation and efficient parallelism of computation with strong spatial and temporal correlation. Our solution is based on a tabularized representation of generic watershed hydraulic processes, which essentially consists of key points in the parameter-state space for watersheds that have been simulated. It is built from scratch during simulations, and dynamically updated as new simulation results are generated. The tabularization is maintained in the database, serving the majority of user queries either directly or through interpolation. To build and maintain the tabularization efficiently, we employ a compute cluster, which features highly parallelized simulation and a fine-tuned key point caching mechanism that effectively reduces redundant computation and the number of new key points to be inserted to the database. Our mechanism only assumes deterministic mathematic models, thus can be expected to be applicable to other similar scientific computation problems.

The remainder of the paper is organized as follows. In Section 2, we briefly review the previous work. In section 3, we give a brief introduction of Hydro-Earth project and define the models of hydraulic simulation considered in this paper. In section 4, we present our caching mechanism to solve the optimization problem. In section 5, our performance results are presented. Finally, we conclude the paper and briefly discuss ongoing and future work in Section 6.

2 Related Work

The canonical performance enhancer to scientific computing has been parallelization, along with from algorithmic improvements. Although parallelization of watershed simulation is a key component to the solution, our main contribution is most pertinent to computation saving strategies explored in computational chemistry.

First introduced by Pope [4], the In Situ Tabulation (ISAT) approach has been widely adopted for function approximation in the scientific community. For the combustion simulation, ISAT use a tabulation method for a particular reaction flow; build up a database table through flow calculation. A major limitation of ISAT is the high storage requirement. Hedengren et al. [1] investigated combining model reduction with computational reduction to handle storage problem. In the DOLFA work [6] by Veljkovic et al., the most performance improvement has been observed with dynamic cleaning of database based on usage statistics. In comparison, our work maintains the database records through careful insertion, avoiding the burden of periodic cleaning. Work by Panda et al. [2][3] have studied the use of high-dimensional index structures to efficiently manage approximation functions and to speed up computation. Veljkovic et al. proposed a modified algorithm for coordinating the building and the distributed management of an online scientific database [5] which proved to be an effective approach to improving efficiency of simulations.

3 Large-Scale Watershed Simulation

Hydro-Earth is expected to handle two typical types of queries. First, history replays and simulation experiments, which specify a geographical region (along with the associated modeling parameters), an extended period (or a spot in time), and (time series of) inputs. The expected outputs are various (time series of) system states for the corresponding time span. Second, continuous queries, which dictate a region and demand results to be produced as new system inputs are becoming available. Both can be reduced to evaluations of the underlying hydraulic models given values of system inputs and parameters. Our objective is reduce their response and execution time by maximally reuse prior simulation results.

3.1 System Architecture

Hydro-Earth employs a three-tier architecture (Figure 1), where a Web application server accepts user requests for virtual environment navigation and location-dependent queries, retrieves relevant data (e.g., satellite images, geographic and

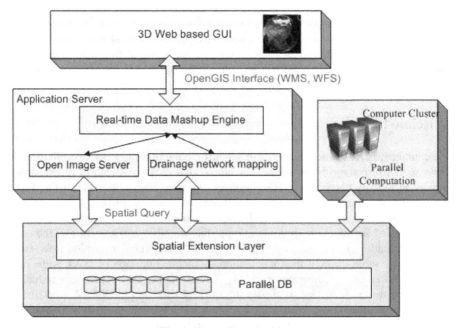

Fig. 1. Hydro-Earth Architecture

hydrologic data of the specified region) from backend databases . The data are being superposed by data mash-up engine and then delivers to Web based client.

3.2 Hydraulic Simulation Models

To calculate the amount of water (calculation of sand proceeds in a similar way) carried by an arbitrary river segment x, for any time step t, computation proceeds in two phases. The "vertical" phase computes $Q'_x[t]$, the amount of water injected locally into x, by simulating local processes of evaporation, water interception, penetration at different levels of soil over the past time interval. Then the "horizontal" phase calculates $Q_x[t]$, the amount of water flowing through x, by incorporating water amounts transferring from upstream segments of x, using the diffusion wave equation. Eq (1) is the Muskingum solution to the equation, where $Q_{x-1}[t]$ and $Q_x[t-1]$ are the sums of water flowing from the upstream segments of x, at time steps t and (t−1). C_1, C_2 and C_3 are coefficients normally dependent on Q.

$$Q_x[t] = C_1 Q_{x-1}[t-1] + C_2 Q_{x-1}[t] + C_3 Q_x[t-1] + Q'_x[t] \tag{1}$$

The simulation is parallelized among a cluster of computers, with a central database handling result persistence and intermediate data exchange. However, a critical performance issue is, throughout the simulation, the same equations are solved numerically at each time step for each river segment, although may be applied with different parameter values. Since a modestly detailed modeling of watershed over a small geographic region may involve thousands of river segments, a minute-level simulation for a time period of a few hours may easily run into millions of such iterations. Such repetition not only results in redundant computation, but also redundant intermediate data transfers, posing strongly negative effects on real-timeliness. We thus expect to greatly enhance the computation performance of the entire process, by eliminating repetitive computation.

4 Efficient Tabularization with Caching

To greatly improve computation efficiency, we aim to: (1)reduce repetitive calculation by building tabularized model, which is built by dedicated compute cluster; (2)achieve efficient parallel computation within the cluster, by exploiting both temporal and spatial parallelism, in both the initial building phase and the incremental building phase of the tabularized model, in order to minimize communication between database and compute cluster and least response time; (3)develop database UDFs(user define function) that produce interpolated results from the tabularized model, given input/parameters specified in user query predicates. We thus center on the fundamental computer science technique of memorization. With memoization, when a function is evaluated on certain arguments, the returned value is cached; whenever the function is called again on the same arguments, the value can be retrieved from the cache. This potentially saves a great deal of computation, as the function does not have to be evaluated repeatedly.

4.1 Algorithm Description

For our specific watershed simulation problem, along with each parameters combination φ that we store in the database, we store the corresponding computed function
value qout with φ as input in the database. Those time-varying coefficients and output
variable constitutes a database table, which we call a "function table", which grows
through computation. Whenever possible, we evaluate function values, leverage linear
interpolation of points from the function table. We thus expect the function table to
steadily grow fuller, as the central copy supporting memorization. For enhanced performance, we further let each computing node to locally cache a portion of the function table. Through the table look-up and interpolation of the retrieved points, each
node could avoid solving the original nonlinear equations as much as possible. A
high-level description of our basic algorithm is given as follows:

```
for each river segment
            Fetch relevant function table values from local cache
and database if not cached
for each point in the simulation sequence
        Range query through the fetched values
        if there exist reference points satisfying
```

$$\left| \frac{x_i - x}{x} \right| \le \varepsilon \Leftrightarrow -\varepsilon \le \frac{x_i - x}{x} \le \varepsilon \Leftrightarrow (1-\varepsilon)x \le x_i \le (1+\varepsilon)x$$

```
        // xi is the current point
        Compute output values by interpolation of the reference
points:
```

$$qout_i = \frac{\displaystyle\sum_{i=1}^{n}\left[qout_i \times \frac{1}{\sqrt{(Seglength_i - Seglength_t)^2 + \ldots + (q2_i - q2_t)^2}} \right]}{\displaystyle\sum_{i=1}^{n}\left[\frac{1}{\sqrt{(Seglength_i - Seglength_t)^2 + \ldots + (q2_i - q2_t)^2}} \right]}$$

```
    else
        Compute output values through the original procedure
        Send the obtained new points to master node
```

4.2 Model Tabularization

Columns in our function table are input/output of a physical procedure. We apply
various normalizations to reduce the size of the function table, for example, when
some parameters, although evidently impact the process, are essentially static for a
generic entity.

We combine the "vertical" and "horizontal" phases of computation on a segment x into an "ensemble of math models", and treat it as a finite state machine with time-varying and time-invariant parameters. Primary variables in diffusion wave equations, are taken as state variables, each combination of external inputs serves as a condition. The operation of deriving all kinds of time series thus becomes an iterative procedure of identifying current state and condition, updating model parameters, looking up the table for the new state, computing other outputs, then updating current state and condition for a new round of table look up.

4.3 Parallel Computation

In a topological sense, we model a watershed as a tree, and parallelize simulation over the tree among the compute cluster using MPICH-2. One node serves has the master node, sub-dividing and dispatching compute tasks to the rest nodes that are slaves. Concurrent slaves each process a different subtree in the river topology. To reduce inserts to the function table, we let the master node determine which new points to add to function table, by grouping very close nodes reported from the slaves. A high-level description of the process within master/slave nodes is given as follows:

```
//Master node
Begin with the first river segment at the lowest level of the tree
while root is not reached
        Seek downstream (up the tree) until a preferred size is reached
        Find bounding box of inputs/parameters for this subtree
        Send coordinates of the tree, along with corresponding portion of
index to next available slave
        Get to the first un-assigned segment at the lowest level of the
entire tree


//Slave node:
while not the last point in the input/parameter series
    if the point is covered by the index cache
        skip
    else
        compute through the original procedure
Send to master new key points, along with results on the connecting edge
to its parent subtree
```

4.4 Reduce Database Access Overhead

As the table size increases, the number of pre-fetched points could grow up. Thus, the execution time of each interpolation step will increase. We reduce table insertion by carefully select inserting points through gradient threshold test, and also put reference points in the memory space using kd-tree. Through these methods, the inserted records decreased significantly. Thus, database access overhead, including insertion and retrieval operation, is greatly reduced.

5 Experiments

We have implemented the afore-described solution in the Hydro-Earth hydraulic simulator. Tests are performed on a cluster of 4 nodes (1 master + 3 slave nodes), each has a 3.8GHz Intel 4-core Xeon CPU and 10GB of main memory, and a dedicated database node installed with Oracle 10g.

We choose "Cache Hit" which is an indicator of the utilization of function table at slave nodes, as a key metric to evaluate our solution. Fig 3 shows results from a typical simulation task, which simulates the hydraulic processes of a regional watershed with 4437 river segments, for 33 simulation days. The highest cache hit reaches 59.4 %, which means there are large recurring intermediate results during computation process.

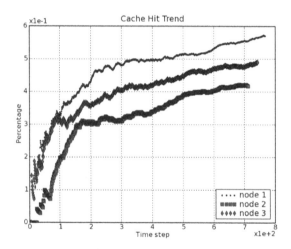

Fig. 3. Cache Hit Trend

We also observed a decrease of 84.1% in table size after applying various key point selection criteria at the compute cluster, resulting in 42,674 records in the function table after running that task. The result proves that through our caching mechanism the redundant computations could be reduced effectively, thus the computational efficiency could also be improved greatly.

6 Conclusions

Leveraging model tabularization, we have designed a fine-tuned caching mechanism to effectively reduce redundant computations for watershed simulation. This method could be easily applied to other scientific computation problems, and is especially applicable to problems with highly involving mathematical models as solving. If the mathematical models are more amenable to analysis, it would be a viable option to systematically deduce simplifying models (possibly for different regions of the parameter space of interest), rather than statistically learning them from simulations. Learning with no a priori knowledge of the procedure is equivalent to black box system identification, which normally has to run constantly and may still render a varying model.

References

1. Hedengren, J., Edgar, T.: In Situ Adaptive Tabulation for Real-Time Control, 3 (2004)
2. Panda, B., Riedewald, M., Gehrke, J., et al.: High-Speed Function Approximation (2005)
3. Panda, B., Riedewald, M., Pope, S.B., et al.: Indexing for Function Approximation, pp. 523–534 (2006)
4. Pope, S.: Computationally Efficient Implementation of Combustion Chemistry using in Situ Adaptive Tabulation. Combustion Theory and Modeling 1, 41–63 (1997)
5. Veljkovic, I., Plassmann, P.E.: A Scalable Scientific Database for Chemistry Calculations in Reacting Flow Simulations. In: Yang, L.T., Rana, O.F., Di Martino, B., Dongarra, J. (eds.) HPCC 2005. LNCS, vol. 3726, pp. 948–957. Springer, Heidelberg (2005)
6. Veljkovic, I., Plassmann, P.E., Haworth, D.C.: A Scientific on-Line Database for Efficient Function Approximation. LNCS, pp. 643–653. Springer, Heidelberg (2003)

The Applications of Mobile Payment

Xin Chen

Electronic Commerce Department
Guangdong University of Foreign Studies

Abstract. With the further development of the Internet and communication technologies, in particular, when 3G technology are widely used, mobile payment is the development trend of mobile e-commerce, there will be a very large commercial prospects and will lead the mobile e-commerce and wireless financial development. This article is to review the mobile payment (asl called m-payment) in different countries.

Keywords: m-commerce, mobile payment, mobile phone application.

1 The Applications of Mobile Payment in India

1.1 India Marching towards M-Commerce

March 24, 2008, India is marching towards m-commerce, a world where you can make all payments by keying in instructions on your mobile phone. In developed countries (particularly Europe) it is common to find people paying for even basic purchases at shopping malls or train tickets with a few clicks of the keypad. India may not be there yet, but we have started the journey [1].

The Bangalore-based company provides mobile-based security and payment services, enabling users to transact, through their mobiles, the payment of utility bills and insurance policies, buying airline, bus and movie tickets, and recharging their prepa id mobile accounts, among other services.

Mobile airline ticket purchasing began in Japan as early as 2000, moved into China in 2006 with Smart Pay, and is now showing promise in India, with Jet Airways' Jet Wallet and Kingfisher's Fly Buy SMS both having launched mobile booking and payment solutions within the last few months. Now that Cell Trust has provided a Secure SMS solution, which inherently provides two-factor authentication, the mobile phone being in the possession of the user and a secure PIN being received over a secure SMS mobile channel, mobile payment has just become much more robust and viable. Until now, the security piece has been a big part of the challenge. With Cell Trust's Secure SMS technology, mobile airline payment may be the 'silver bullet' for which the industry has been looking Cell Trust's Secure SMS Gateway manages and exchanges messages via its highly Secure Mobile Information Management platform, giving businesses and government the ability to confirm that the recipient's mobile device is able to receive secure SMS messages, easily market and push the Cell Trust Secure SMS micro-client to their customers' mobile phones so they can receive and send secure text messages instantly; use the same set of APIs to send both secure and

Y. Wu and Q. Luo (Eds.): ICHCC-ICTMF 2009, CCIS 66, pp. 62–67, 2010.

normal SMS or text messages; and definitively measure ROI of mobile programs and their effectiveness [1].

1.2 Barclays Bank's "Hello Money"

Barclays Bank recently launched 'Hello Money', a mobile banking service that enables one to do an entire gamut of things, enquiry, funds transfer, bill payments and requests for financial statements. Until this product arrived, the best 'mobile banking' one could do was to check bank balance or phone-in instructions to the banker. But with Hello Money, a customer of Barclays can, for example, pay his electricity bills via the mobile phone. Click a few keys and the account is debited, the bill is paid, all without any human interface. Over time, Barclays intends to enlarge the scope of this service to include payments to a whole lot of other things, such as for purchase of an air ticket [2].

Unstructured Supplementary Service Data (USSD), it is a technology unique to GSM mobile services and offers information transmission over existing networks. This means users do not have to enable a GPRS link on their phone (as is done currently) to download any application Also, USSD provides session-based communication, which means the service does not store any data or banking session on the handset that can be misused by a third party. There are no records of the transaction one completed, no SMS updates that are stored (a common feature in other banks' mobile banking solutions).

As all operations happen in real time, data used is not stored in the handset memory or the GSM network but only in the Barclays Bank server.

Consequently, turnaround response times for these applications are shorter than SMS. The primary benefit of USSD is that it allows for very fast communication between the user and an application. Most of the applications enabled by USSD are menu-based and include services such as mobile prepay and chat.

For the user this means banking is now available on a toll free number (*598*1#) that will guide one to a menu-based system. All handsets, low-end to high end, can access this service at a monthly subscription of Rs 30, just like how ring tones are subscribed to. The service is available on roaming at no extra charge [1].

1.3 mChek in India

1.3.1 The Challenges

Awareness and adoption will be a key challenge. Another challenge will be to bring the eco-system of banks, telecom operators, payment industry players and merchants onto a common platform so that consumers can realize the value offered by mobile payment platforms.

The mobile phone can become our primary interaction/transaction tool and can eventually be the primary network access device.

A host of network-based services, including information services, social networking services, identity management solutions and secure transaction-based services, can eventually move to the mobile platform [4].

1.3.2 The Growth Drivers

One of the key factors that will drive adoption for mCommerce is that now more than 290 million Indians have a post-paid bill to pay or a pre-paid top-up to be made every month.

Even at a 3 per cent conversion, the industry would be around Rs 3,000 crore. The ability to provide a simple, secure solution that works on all mobile phones will be a key factor in driving growth.

We believe that once the habit of making payments to mobile service providers is made, the expansion of this set to other payments and purchases will only be a matter of time.

1.3.3 The Revenue Model

mChek works on a hosted application service provider (ASP) model and generates revenue from clients and merchants based on transactions completed on mChek [3].

2 Other Aspects of Mobile E-Commerce Application

2.1 Sybase 365 Improve Mobile Payment

Sybase 365, a subsidiary of Sybase, a provider of mobile messaging services, announced its commitment to a global mCommerce strategy.

The company stated that until now, mCommerce has been slow to take off due to issues such as technology constraints, local regulations and most importantly the lack of cooperation between mobile operators and banks in individual countries and around the world.

For the past several years, Sybase 365 has been providing remote mobile payments via SMS, MMS and WAP to content providers and brands, charged directly to the subscribers bill via their mobile operator. This, in effect, is mCommerce paying for digital mobile content and services delivered via the mobile phone. The vision is that other digital and physical goods and services will be available for purchase over the operator's network, through the traditional payment channels, but also by credit card, direct debit from Bank accounts, stored value or even by using the mobile as a virtual wallet swiping it at a point of sale terminal using NFC technology [5].

Sybase 365 has consistently been making inroads in the mCommerce arena. In October, 2007 the company launched Sybase mBanking 365 a suite of products for the financial industry enabling banks to interact with customers in real-time through mobile alerts, two-way banking services, out-of-band authentication, and marketing campaigns. The solution also offers a natural language capability Answers 365 enabling consumers to interact with their bank using their own words.

In January 2008, Sybase 365 announced its partnership with C- SAM, a provider in secure mobile phone-based transaction technology, to extend its mBanking solution to include a downloadable Java and NFC compatible client. Banks worldwide can now offer mobile banking services through SMS, WAP, and a downloadable client, providing increased flexibility to interact with their customers all through a single connection to Sybase 365 [5].

2.2 Blaze Mobile Wallet

2009 Feb 28 - (VerticalNews.com) - Blaze Mobile [7] announced a version of its Blaze Mobile Wallet, the first fully- integrated, secure mobile commerce solution for handheld devices, that supports iPhone and iPod touch is now available on the Apple App Store. The Blaze Mobile Wallet gives users the ability to quickly transact everyday purchases and easily manage personal finance tasks like banking and fund transfers.

The Blaze Mobile Wallet application is available for a one-time download fee of $1.99 from Apple's App Store on iPhone and iPod touch or at www.itunes.com/appstore/ About Blaze Mobile Blaze Mobile develops innovative mobile commerce and advertising solutions that enable secure, convenient, cost-effective transactions and promotions from the mobile device. The Blaze Mobile Wallet enables fast and easy "contact-less" purchases -- including movie and event tickets -- as well as fund transfer, banking, and personal finance management. It can manage bank accounts at more than 8,000 supported financial institutions, view electronic receipts, and quickly and easily create expense reports. Blaze Mobile Wallet also offers valuable location based services such as maps and points of interest, including ATMs, restaurants and more - all in the palm of your hand. The Blaze Mobile Advertising Network enables companies to target specific mobile advertisements and promotions to their most important customers, delivering unparalleled marketing and branding capabilities.

2.3 Real-Time Financial Decision Support

The financial decision support process in a mobile environment is a dynamic process that evolves with different context changes, is characterized by fluctuating uncertainty, and depends on multi-attribute preferences of the individual mobile decision maker. It should link past, current and future states of the mobile environment and needs to be adaptable to user and system constraints. In addition, mobile decision support requires the underlying distributed computing infrastructure with wireless and mobile networks as the main components).

These are some issues that the mobile consumer might be considering when deciding which payment option is best to minimize transaction fees, minimize credit card debts, and maximize monthly savings. Currently, many such decisions are based on intuition or past experience, and there are no analytical tools developed to assist mobile decision-makers in these situations. There are a few products on the market that provide different personal finance management solutions [6].

Some of these products are Microsoft Money, Smart Money, My Money, Quicken, and Mind Your Own Business (MYOB). These products support and manage operational transactions and permit some analysis of historical data. MYOB, for example is aimed specifically at Australian small-business users. MYOB replaces "the cash register with a point-of-sale system that streamlines store operations and manage sales, stock, goods and services tax (GST), staff and customers". Microsoft Money allows a consumer to view vital financial statistics at a glance, set up regular bill payments as reminders, or have them taken from an account automatically, plan and maintain budgets, and see projected cash flow so the user knows how much to spend. However, most of these are not yet customized for access from a mobile device. The only mobile application that comes close to the decision problem described above is Microsoft Money for Pocket PC (http://www.microsoft.com). This mobile version of Microsoft Money, however, provides little decision support as it only allows the user to view the account balances (upon synchronization with desktop Microsoft Money) and record any new transactions.

2.4 Mobile Accounts Management

A mobile accounts management problem refers to a mobile user's problem of selecting the best possible immediate payment option in relation to the associated future gains or losses. Payment options in an e-commerce context include payment by electronic funds transfer at point of sale (EFTPOS), electronic wallets, electronic coupons or electronic cash, and of course, more-traditional options (for example, paying by cash, lay-by, credit card, or cheque). With so many options to choose from, the mobile user can enjoy the convenience of paying for products or services electronically, and charging each to one of multiple bank accounts (savings, credit, cheque accounts). Some online financial services (such as online banking) allow the consumer to view the account balances online, or to record transactions electronically to keep track of his/her transactions and balances. A tool that helps the consumer save money and/or manage their accounts efficiently is regarded as highly attractive to the mobile user. [7]

Our approach of supporting the mobile user with QoD indicator is also relevant to the mobile accounts management problem. This indicator can be useful to the mobile consumer when making real-time decisions about efficient accounts management since, in particular, together with the recommendation on the best option to realize future gains, they are informed of the QoD that was used to calculate the choice, or be alerted when QoD level was not good enough to justify the suggestion. [8]

The application area of mobile accounts management is described as a means for illustrating how such a framework could be utilized. A prototype system, iAccountsMgr is designed to provide mobile decision support for this area. The proposed procedure aims at providing a mobile user with on-the-spot assistance for decision making concerning the payment method for products and services aiming at efficient management of the periodic (monthly) budget. Equipped with the proposed system, the user will be able to assess the future consequences of a choice, and alerted about the implications when buying an item on the move, or before charging emergency purchases against her bank accounts. The main innovation of the mobile DSS is that together with calculating possible scenarios the system provides a measure of reliability – the QoD – to each scenario according to the data used in calculating these scenarios. We feel that combining strategic information related to the decision at hand with a measure of the quality of the information (the QoD indicator) provides the decision maker with the ability to make a better informed choice based on the most up-to-date, relevant information available. [9]

3 Views and Prediction about Mobile Payment in China

Recently Tom online and Union Mobile Pay Ltd (which is under the umbrella of China Mobile) had signed a strategic alliance agreement on cooperation of mobile-pay. This enables their users to pay airplane tickets, lotto, communication fees and even their bills of electricity and water by mobile phone.

According to Analysys International's study(Analysys International is the leading Internet based provider of business information about technology, media and the telecom industry in China with the mission to help their clients make better business decisions.) ,mobile-pay will become a safer and faster payment method than online payment. It predicts that the user number will reach 18.7 million by the end of year 2005, with a dramatic annual growth rate of 1068.75%; and the penetration rate will reach 4.74%. By the end of year 2008, the total user numbers will reach 97.5 million. Analysis International also points out that the current mobile payment market is still under an introduction phase of user cultivation and gives strategies on business models and value chain development.

The current mobile payment market dominated by operators is still under an introduction phase of user cultivation. For operators, services of mobile payment should be regarded as vertical complementary services to their various technical fields rather than expansions to their system platform services. On the other hand, the industry value chain hasn't been established. Players in the value chain should enhance their cooperation and coordination to build up a health and complete cooperation system and platform for the value chain and drive the development of the mobile payment applications.

Analysys International thinks that although the business model of mobile payment is still unclear, players have shown their passion in innovation. The business model will be clearer and cooperation model will get more support from the industry authorities."

References

1. Pay through the mobile phone, Businessline. Chennai, October 6 (2008)
2. Analysys International Predicts That the Mobile-Payment Users Will Reach 97.5 Million by 2008, PR Newswire, New York, November 9 (2005)
3. Bank on your mobile, Businessline. Chennai, March 24 (2008)
4. Cell Trust Preps Airlines On M-Commerce, Wireless Business Forecast. Potomac, April 18 (2008)
5. Sybase 365 Ramps Up with m-Commerce strategy, Wireless News. Coventry, February 14 (2008)
6. Financial Joint Venture May Finally Kick start M-Commerce, Wireless Business Forecast. Potomac, March 6 (2008)
7. Blaze Mobile; Blaze Mobile Wallet Now Available on Apple App Store, Investment Weekly News. Atlanta, February 28 (2009)
8. Burstein, F.: Support for Real-Time Decision Making in Mobile Financial Applications, ch. 41
9. [America] Deak, E.J.: The Economics of E-Commerce and The Internet (November 2006)

The Challenges and Strategies of Commercial Bank in Developing E-banking Business

Xin Chen

Electronic Commerce Department
Guangdong University of Foreign Studies

Abstract. Electronic-banking (e-banking) businesses mostly are developed based on traditional business of commercial banks. The normal method is combined traditional business with internet in order to realize new E-banking service.The problems existing by doing so are: lack of effective measure to control commercial risk with single technical solution, lack of laws of E-banking, simple methods of marketing, service and training for E-banking business; weak construction of bank's infrastructure setup. All above-said problems have limited the development of E-banking business of commercial banks. In this article we will discuss the e-banking evolution and explain the emerging of e-banking services, E-payment system as well as legal, risk management challenges, and have presented necessary suggestions for managing e-banking.

Keywords: Commercial Bank, E-banking Business, E-banking, strategies.

1 Introduction

1980s, commercial banking has continuously innovated through technology-enhanced products and services, such as multi-function ATM, tele-banking, electronic transfers, and electronic cash cards. Over the past decade, the Internet has clearly played a critical role in providing online services and giving rise to a completely new channel. In the Internet age, the extension of commercial banking to the cyberspace is an inevitable development [1]. Both researchers and practitioners in the BFI have highlighted the need for banks to broaden their branch-based delivery channels by embracing electronic banking (e-banking). [2]

This article will discuss the The Challenges of Commercial Bank in Developing E-banking Business.The article points out commercial banks of China should walk out a way that is suitable for current stage of E-banking business development and current economic situation. Although the development of E-banking business faces many problems, especially some institutional and operational ones,that can not be solved in short term,commercial banks should treat them carmly.No rush and no passive. [3]

2 Current Development of Commercial Bbanks in E-banking Business

Banking has evolved over the past few years from a face to face (customer to teller) service to that of "anytime/anywhere/anyhow" banking services, using various delivery channels. E-banking is the way forward for the banking industry. [4]

Y. Wu and Q. Luo (Eds.): ICHCC-ICTMF 2009, CCIS 66, pp. 68–74, 2010.
© Springer-Verlag Berlin Heidelberg 2010

Traditional banking must be relied upon to continue to earn money, while financial innovations over the Internet draw in new accounts and enable the e-bank to tap into a wider customer base and create new opportunities in ending and funding. [5] Banks maintain online trading systems that can be accessed 24 hours a day by any participant around the world. Time zones do not matter and communication is both richer and faster. [6]

In essence, traditional banking is a way of person-to-person service delivery over the counter. It views the ustomer as a passive participant in the transaction process until the time of the transaction. Conversely, e-banking ervices are promising customized services tailored to customer needs. Such services delivery considers the ustomer as an active participant at every stage of the transaction process, and as the co-producer of the financial roducts and service offerings. In contrast to the traditional banking services which focus on articulated needs as tated by customers, or the bank's perceptions of customer demands, customized services focus on both articulated nd unarticulated needs by guiding customer's profile andusage patterns. Instead of accepting off-the-self financial roducts or services, customers can choose their personalized financial services in an e-banking context. However, e-banking offers alternative approaches by which banks can provide individual offerings and services to attract ustomer interests, increase customer loyalty, and repeat transaction [7].

In addition, e-banking can be utilized as a channel to develop long-term customer relationships through ready access to a broad and increasing array of products, services and low-cost financial shopping, rapid response to customer inquires, and customized product-service innovation [1].

Consequently, the alternative channel may extend the bricks-and-mortar banking. In sum, e-banking realizes the following value propositions: efficiency, convenience, customization, and market extension [8].

3   The Challenges and Problems of E-banking

3.1   Security

Given the open nature of the Internet, transaction security is likely to emerge as the biggest concern among the e-bank's (actual and potential) account holders. Since transactions risk would create a significant barrier to market acceptance, its management and control are crucial for business reputation and the promotion of consumer confidence as well as operational efficiency. Before launching new products and services, it is therefore imperative for the e-bank to implement measures to safeguard client assets and information (for example, a confidential database with firewalls and stringent access control), and to advertise as widely as possible the introduction and expert endorsement of initiatives to maximize transaction security. [5]

3.2   Dynamic Environment

To successfully cope with the challenge of the e-banking innovation, the incumbent banks must understand the nature of the change and capability barriers that it presents [9]. Without this understanding, attempts to migrate to e-banking may be doomed to failure. Banks that are equipped with a good grasp of the e-banking phenomenon will be more able to make informed decisions on how to transform them into e-banks and to exploit the e-banking to survive in the new economy [9]. Given the e-banking is a financial innovation [1]; the change may render the organizational capabilities of the traditional banks obsolete. From the resource-based view [10], in such a context, the banks must constantly reconfigure, renew, or gain organizational capabilities and resources to meet the demands of the dynamic environment. Developing core capabilities can help the banks redeploy their resources and renew their competences to sustain competitive advantages and to achieve congruence with the shifting business environment. [2]

3.3 Service

E-banking enters this New Year in a condition that, if not anemic, surely isn't robust. While almost all of the large banks offer it, few of the many small banks do-or even plan to. On the retail side, banks are offering credible products, but most consumers still aren't turning on to the value propositions. Oil the corporate side, the prospects is brighter, but here it's much harder to deliver a comprehensive product. Common technology supports both markets.

As a card-carrying early adopter, I call report now, as I did last year, that for file the product works as advertised, and I wouldn't consider do ing without it. I've tried three internet-only e-banks and as many traditional banks that offer e-banking. Two of the banks performed so poor ty that I discontinued using them. Others just faded away, or I gravityed to a better service. Here are same of my gripes.

The consumer never sees the equivalent of a canceled check to prove that a bill payment was actually received, even though I see when it was deducted from my account. One of my payments was suspended in limbo for six months while "pay up or else" notices piled up.

About half of all "electronic" bill payments end up as paper checks sent by the bill-pay provider.

Electronic bill presentment won't help much because low-volume billers won't do it. One of my e-banks displays pending payments on a different screen from those already sent out. I would have had to manually add up the pendings and deduct that total from my actual balance. That's unacceptable. I switched.

I was shocked to learn that some e-banks do not accept or initiate one-time electronic deposits, though of course they must accept regular monthly direct deposits. So in some cases I have to send paper-check deposits via regular post, never sure when they will arrive and become available.

In one instance, an e-mail query was answered by e-mail-which gave us the telephone number of a customer service department. The same institution batted my query about ACH payments between the bank and the e-banking service provider because they weren't sure who was responsibLe.

The clear message I get from these trivial-seeming hassles is that the industry still isn't serious about providing a reliabLe, user-friendly service. That's annoying to early adopters and inhibiting to those in the still-underwhelmed mass market. A huge fraction of consumers who sign up for e-banking aren't committed users; many others quit because they're disappointed in the service. [11]

3.4 Standards

If there is one root problem for the lackluster performance of e-banking, I think it is the industry's failure to attend to that banking basic, the payments infrastructure. Time and again, interesting, soundly conceived, impressively backed innovations are proposed, beta tested and piloted, but never fully embraced by the industry. Bankers and observers such as me have hailed the Launching of middleware platforms OFX, the Gold standard, and IFX-but we still don't have a single industry standard. XML is now emerging as an all-purpose, platform-independent Language, but banks' slow pace in adopting it may imperil its future success. The Spectrum switch is a well-conceived solution to the authentication problems of bill payment systems. The New York Clearing House has launched a universal payment identification coding system that could speed adoption of bill presentment. A FleetBoston spin-off, Clareon, offers a well-crafted B2B payment system powered by e-mail.Citibank came up with C2it, a universal system that executes free funds transfers between the banks of any two parties, even if neither of them has a Citi account.

Online Resources has long offered an e-banking system that features transferring funds in real time over ATM networks directly into its proprietary billpay and e-banking engines.Clearly, there is no shortage of good deas. But more innovation is not what the industry needs right now. Strong, profitable growth won't be assured until the industry gets behind standards assuring that any bank can deliver a competent, reliable service without extra fees. [11]

4 The Strategies of the Development of E-banking

Philipp says that to succeed, e-banks must stay away from mass marketing with its massive costs. Don't build the bank around depository accounts, but around a mix of standard financial services plus wealth management, financial planning, and online lending. Go for owning a big share of a targeted niche market, she advises. [11] Consequently, the new IT-infrastructure should ensure the interoperability and transparency in addition to covering the requirements for security. [2]

4.1 Planning New IT-Infrastructure

Most innovations in financial services have been enabled by the innovative application of IT [12]. E-banking initiatives are based on the Internet and require ntegration with existing systems. For the traditional banks, the need for designing new IT-infrastructure and developing appropriate technical platform may be self-evident. In addition, e-banking adds some subtleties to system integration capabilities, particularly with regard to network and platform integration. Hence, it is imperative to evaluate enabling and emerging IT, to upgrade network architecture, to erect open platforms and to integrate the existing applications with Internet. Banks must also know how to develop new solutions to integrate internal systems with external business stakeholders, allowing them to conveniently carry out secure transaction online. Consequently, the new IT-infrastructure should ensure the interoperability and transparency in addition to covering his requirements for security. [2]

4.2 Enhancing Transaction Security

Security is a significant determinant of willingness to use banking services [1]. Given the open nature of the Internet, cyber-crime will increase in e-banking environment. Online transaction security and privacy protection are likely to emerge as the greatest concerns among the financial institutions and consumers. Since transaction risk would create a great barrier to e-banking acceptance, its prevention and control are crucial for bank reputation and the promotion of customer trust. Before launching new e-banking products and services, it is therefore crucial for the bank to implement safeguards embedded in the IT-infrastructure to manage client assets and private information, and to advertise as widely as possible the introduction and expert endorsement of transaction security initiatives [13]. There are imperatives of enforcing appropriate security policy and procedures within the e-banking processes. Here, public policy and governmental regulation of online transactions will strongly influence the development of security policy of e-banking. Thus, understanding the financial regulations of different countries is critical for banks to develop their security policy. However, the success of e-banking is likely to come with the ability to developing sound solutions of privacy and security [1, 2].

4.3 Providing Value-Added Content

Adoption of e-banking relies on effective information exchange between banks and their customers. When e-banking replaces a traditional service, the customer will need real-time access to elevant information to reduce the uncertainty of transaction compared to the person-to-person services in a branch office. Banks have to give up its information asymmetry relative to its customers to enable them to perform self-services [7].

4.4 Delivering Differentiated Services

A firm's competitive advantages generally come from providin ifferentiated products and services [14]. Automation of banking services is expected to reduce the need or standard services while there will be continued demand for differentiated services [15]. Therefore, in his transformation to e-banking, another challenge for banks is how to differentiate their banking services from ther banks. For example, bundling and cross-selling are possible strategies for differentiation [16]. Another strategy is service customization which offers tailor-made individual offerings and services to the clients.The customized service is a combination of the customization of the operations and the customization of the marketing and customer relationship [7]. It offers customers more control in the transaction process and targets to solve the particular needs. Banks can use data mining techniques to analyze the customers' patterns of doing business and preferences, so as to influence customer decision-making by framing the choice options and making it easier, more productive, more engaging, and cheaper for customers to deal with them than with competing banks. Here, success in this endeavor would enhance the differentiation of financial services, which would in turn rengthen customer satisfaction and loyalty, increase repeated purchases, and attract new customers [17].

4.5 Conveying Value Propositions

The banks should use e-banking to focus on customer needs in order to gain the strongest competitive advantages [7]. Thus, it raises a fundamental question whether the e-banking channel offers new value to the banking activities of the customer base it serves. As discussed previously, e-banking has its own inherent benefits, especially efficiency, convenience, service customization, market extension, etc. Banks must have the specific capability to translate the potential benefits into actual value propositions. This capability refers to the degree to which the banks are able to envision customer's expectations, take new value propositions to market, and educate the existing and potential customers [18]. As far as customer's expectations are concerned, Liao and Cheung [1] found that individual expectations regarding accuracy, security, transactions speed, user friendliness, user involvement, and continence are the most important quality factors in the perceived usefulness of e-banking. In addition, the customer's acceptance of e-bankinggreatly depends upon their experience with existing financial services and with enabling technology. Therefore, educating the customers to enhance their understanding of e-banking would increase their willingness to accept it.

4.6 Managing Customer Relationships

The customer relationships are the source not of temporary gains but long-term profits [19]. A critical success factor for e-banking is how well it executes customer relationships management between banks and existing/potential customers. Since demographic change is an important issue in evolution of e-banking, to exploit the change and increase market share, banks must seek to attract and capture such potential clients as early as possible by supplying a low switch cost technologically and innovative and sophisticated products and services such as global e-banking and mobile transactions over the Internet. The capabilities to accurately identify customer segments and estimate their profit potentials and then target segments that produce high profits and low risks are important for banks. Furthermore, the retention and expansion of relationships with old and lower IT awareness customers are also critical for banks. The optimal marketing strategy for banks is likely to be the cultivation of the demand side along the paths of least resistance. Education and facilitation is a key part of facilitating them to migrate to e-banking. Banks should grasp the full significance of providing detailed information and knowledge to enable the old customers to become sophisticated online clients. However, increased customer sophistication and involvement have a significant effect on the willingness to accept and use new offerings [1, 2]

References

1. Liao, Z., Cheung, M.T.: Challenges to Internet E-banking. Communications of the ACM 46(12), 248–250 (2003)
2. Wu, J.-H.: Core Capabilities for Exploiting Electronic Electronic Banking. Journal of Electronic Commerce Research 7(2) (2006)

3. Gupta, M., Rao, R., Upadhyaya, S.: Electronic Banking and Information Assurance Issues: Survey and Synthesis. Journal of Organizational and End User Computing 16(3), 1, 21 pgs (2004)
4. Sanayei, A.R.: Hamed E-banking Evolution in Third Millennium. Journal of International Marketing & Marketing Research 31(1), 3–13 (2006)
5. Liao, Z., Cheung, M.T.: Challenges to Internet E-Banking. Communications of the ACM 46(12) (December 2003)
6. Felgran, S.D., Novos, I.E., Collardin, M.: Banking: An anatomy of an e-transition. International Tax Review 12(8), 41 (2001)
7. Wind, Y.: The Challenge of Customization in Financial Services. Communications of the ACM 44(5), 39–44 (2001)
8. Stamoulis, D., Kanellis, P., Martakos, D.: An Approach and Model for Assessing The Business Value of E-Banking Distribution Channels: Evaluation as Communication. International Journal of Information Management 22, 247–261 (2002)
9. Simposon, J.: The Impact of The Internet in Banking: Observations and Evidence From Developed and Emerging Markets. Telematics and Infromatics 19, 315–330 (2002)
10. Grant, R.M.: The Resource-Based Theory of Competitive Advantage. California Management Review 33, 547–568 (1991)
11. Orr, B.: American Bankers Association. E-banking: What next? ABA Banking Journal 93(12), 40, 4 pgs. (2001)
12. Holland, C.P., Westwood, J.B.: Product-Market and Technology Strategies in Banking. Communications of the ACM 44(5), 53–57 (2001)
13. Baddeley, M.: Using E-cash in the New Economy: An Economic Analysis of Micropayment System. Journal of Electronic Commerce Research 5(4), 239–253 (2004)
14. Porter, M.E.: Competitive Advantage: Creating and Sustaining Superior Performance. The Free Press, New York (1985)
15. Sannes, R.: Self-Service Banking: Value Creation Models and Information Exchange. Informing Science 4(3) (2001)
16. Altinkermer, K.: Bundling E-Banking Services. Communications of the ACM 44(5), 45–47 (2001)
17. Schaupp, L.C., Belanger, F.: A Conjoint Analysis of Online Consumer Satisfaction. Journal of Electronic Commerce Research 6(2), 95–111 (2005)
18. Wheeler, C.: NEBIC: A Dynamic Capabilities Theory for Assessing Net-Enablement. Information Systems Research 13(2), 125–146 (2002)
19. Nelson, S.: Keeping Old Customers, Finding New Customers and Growing Profits. The Gartner Group (1997)

Easy Domain Processing over Heterogeneous Databases: A Unified Programming Paradigm

Rui Liu, Weihong Wang, Qinghu Li, Tao Yu, Baoyao Zhou, and Cheng Chang

HP Labs China
{liurui,weihong.wang,qinghu.li,tao.yu,baoyao.zhou,
cheng.chang}@hp.com

Abstract. With surging amounts of data becoming available to scientific and engineering professionals each day, demands are growing for (1) extremely efficient data processing mechanisms, and (2) lowered entries to incorporate domain expertise for higher value. With video analytics as an example, the paper pursues a database approach and proposes a unified programming paradigm, which essentially allows one to seamlessly program domain processing workflows, as well as computations to be pushed down to (heterogeneous) database systems, in one programming language. We hence offer great easiness in building efficient data intensive computation applications, with multiple heterogeneous databases as the underlying computation platform.

Keywords: DBMS, programming paradigm, annotation, user defined function.

1 Introduction

Data-intensiveness has become a pressing issue that exacerbates cost of computation. Take a retail chain store video surveillance system as an example, where videos captured in different stores are hosted by different regional data centers through storage services. A typical use case is to *query by example*, *e.g.*, given an image of a customer, the system need to answer in near real time, whether this person is shopping in any store and where exactly he or she is in that store. Answering such a query implies extensive computation over intensive data, including feature extraction from video frames, similarity match on a potentially large distributed data set, *etc*.

To build a scalable and efficient data-intensive computation platform, the commonly accepted norm is to move computation closer to data, avoiding the cost of migrating high volumes of inputs or intermediate results between nodes. In compute clusters, which are the canonical solution to expensive computation, necessities have been raised to strengthen local data management and to reduce data movement, in face of growing data volume [6] [7].

In this paper, we take a different approach and build the data-intensive computation platform around database systems themselves, with the retail chain store video surveillance as an example. Our rationality is based on the following facts. First, a large-scale video processing platform necessarily involves (multiple and very likely heterogeneous) database systems, for managing video contents or their metadata. The data management capabilities are clearly advantages for managing complex

Y. Wu and Q. Luo (Eds.): ICHCC-ICTMF 2009, CCIS 66, pp. 75–82, 2010.

computation tasks. Second, domain processing capabilities of a database system are, to a certain degree, extendable by user-defined functions. Third, video processing, although expensive, produces intermediate results that tend to be reused many times, e.g., feature vectors, and are worthy of persistence and management.

This paper especially focuses on one challenge --- with video data managed by multiple heterogeneous database systems, how can a video analysis professional efficiently program the work flow of video processing tasks without knowing the intricacies of database internals while having video processing computation running inside the database systems?

Our major contributions are a unified programming mechanism and a DBMS-independent runtime, which effectively shield the distinction between the development of application logic and that of database extensions, disguise the intricacies of different database systems, and clear the need for accomplishing video processing workflows using multiple languages.

The rest of the paper is organized as follows. Section 2 discusses related work; Section 3 introduces the unified programming mechanism; Section 4 details the annotation based UDF definition and on-demand deployment of UDFs, with a nearest neighbor query as an example; and Section 5 concludes the paper.

2 Related Work

The Slone Digital Sky Survey work on SQL server cluster has shown dramatic performance gain, as domain computations are optimized by relational query optimizers in the form of SQL UDFs [4][5]. Clustera [3], as an integrated computation and data management system, also demonstrates that a general-purpose cluster can serve as a scalable and efficient platform for very different types of computation tasks, when states of jobs, nodes, logical data units are well managed in a database.

To ease the difficulties of performing large-scale computations, data-parallel application frameworks, such as MapReduce [8], allow users to provide only a few core functions written in high-level programming languages, and automatically schedule and manage their execution according to the implied workflow and parallelization model. Our system further allows the programmer to conveniently program the main application logic and primitive functionalities to be pushed down to database seamlessly in one program using their favorite programming language, and the database query optimizers will schedule the functions inside databases.

Recent language innovations such as LINQ facilitate compositional programming with data in object-oriented languages, however, they are not meant for programming in-database computations. Scripting languages, such as SCOPE [9], Pig Latin [10], and Sawzall [11], allow easy programming of data flows, while none aim to extend functionalities of databases or address their heterogeneity.

In dealing with heterogeneous databases, database federation techniques [12] (e.g., through wrappers) can transparently map data schema and functions of different data sources into one relation database view. However, they do not offer mechanisms to extend the distributed and heterogeneous database systems into a computation platform, which will conveniently retain the data management advantages.

3 Programming Paradigm

For data-intensive computations, the programming style is especially influential to the scalability, efficiency, and robustness of the system, and productivity of users and programmers.

3.1 Application Logic vs. Data Store

There is a traditional delineation between database and applications, which dictates that database systems are data stores that perform (relational) data manipulation at most while applications fulfill business logic. The prevalent programming style, adopted by LINQ, ADO, POJO (Plain old java object) with Java Persistence API, JDBC, IBATIS, *etc.*, is essentially computation-oriented. Due to the lack of mechanisms for pushing down UDF to database, the application performs the entire flow of data processing, and access to database only serves as a way of I/O. This programming style would not work for data intensive computation, due to the impeding cost of loading data from database.

3.2 Programming Workflows vs. Programming UDFs

Data set oriented scripting languages SCOPE and Pig Latin directly instruct processing over data stores, blending a procedural flavor with SQL. They can explicitly express data flows as DAGs (directed acyclic graphs), but are insufficient in expressing algorithms like SIFT feature extraction.

Database user-defined functions (UDF) can manipulate data at various granularities (e.g., per tuple, columns, and rows), and have been commonly adopted to accelerate data access and reduce data transfer [4]. However, UDF implementations are tightly coupled with the database system. Porting them across different types of databases is a very intricate programming task, especially when database specific advantage features are used, e.g., cross-call cache in table-valued functions.

The MapReduce programming model is amenable to both professional programmers and scientific users, whose job boils down to providing two functions (at the minimum), map and reduce. The framework scheduler controls how the functions run on a large number of distributed nodes. This, however, rules out the common database users, who are more comfortable with querying large datasets using SQL. Tasks easily specified by SQL, e.g., build indices or evaluate sophisticated queries, can be very complex for MapReduce programmers.

3.3 A Unified Programming Paradigm

3.3.1 Roles

In face of the above conflicts, we differentiate three types of roles in building and running a large-scale video processing application, where different expertise only focuses on their problems of interest. The first role is the video processing primitives provider, who implements building block algorithms for video processing, e.g., feature extraction, SVM. The second role is the function assembler, whose task is to compose primitives into higher level functionalities. To express the underlying workflow, they can have SQL stored procedures or high level programming languages (Section 4.3) at their disposal. For example, a higher level function can be assembled as the sequence of three primitives: extracting certain types of features from key frames, generating image signatures from those features, and building an index over the signatures. The last role is the end user who simply composes queries in SQL, scripting languages like Pig Latin, or using graphical tools such as a UIMA [1] GUI, which essentially dictate workflows of assembled functionalities (and primitives as well, for advanced users).

3.3.2 Programming Transparency

The two roles of function assembler and primitive designer are often undertaken by the same person, who has domain processing expertise and designs an entire computation flow on the platform. What happens in the common practice would be that this person switches between multiple development environments and different programming languages, to write UDFs and applications, especially when working with heterogeneous databases. We hence offer a unified programming style, with which:

(a) The distinctions between the development of database extensions and that of application level workflows are disguised. The fact that video processing primitives will eventually run in databases is transparent to the primitives provider, who do not need to consider how to extend or deploy new functionalities to database.

(b) A single programming language suffices. We define annotations [17] in high level programming languages, so that a developer can program the primitives and the flow control code seamlessly using one language. We have designed a runtime that automatically extracts the UDFs and deploys them on demand. In the flow control program, stubs for the UDFs are generated as delegates for remote UDFs /SQL calls. Currently, we support Java as the programming language, and it is relatively easy to extend support for other JVM languages like Groovy, Scala and Clojour.

(c) Intricacies of heterogeneous databases are disguised. The developer only needs to code the primitives once, turning the primitives into database specific UDFs will be automatically handled by the system runtime.

4 Video Processing in Multi-databases: A Nearest Neighbor Query Example

We explain the details of user program, UDF deployment and execution runtime, with the nearest neighbor query as an example.

4.1 Similarity Measurement

Similarity between two images is measured as the distance between their feature vectors of a certain agreed-upon type. For robustness, the feature vector is preferably a combination of global and local characteristics, such as [13][14][15]. In the code snippets below, we assume that feature vectors of the images have been precomputed, and stored as attributes to the corresponding images.

4.2 Nearest Neighbor Query in Multiple Databases

To efficiently search for the nearest neighbor from multiple databases, we index feature vectors from all the participating databases with a hierarchical k-means tree [16], where each branching node splits its feature vector set into k clusters, and each leaf node points to a bin of clustered feature vectors. Once the user application issues a query, the runtime goes through the global index, and then sends sub-queries to those databases holding clusters that are closest to the sample feature vector. When sub-queries return, the nearest neighbor will be decided from their respective results.

4.3 Unified Programming

The following code snippets define a *KeyFrame* class, each instance of which is capable of selecting the best matching key frame with itself from a list of key frames.

4.3.1 Object Relation Mapping
We employ the POJO O/R mapping of Java Persistence API to represent relation data as objects. For example, with

```
@Entity
class KeyFrame{
    @ID
    Long ID;
    FeatureVector feature;
    ......
}
```

the class *KeyFrame* is mapped to a database table named *keyframe*, its attributes mapped to table columns. During the runtime, each tuple of the relation corresponds to a Java object.

4.3.2 Operation Mapping
Operations on a relation are defined by methods in the Java class using an annotation, *@UDF*. The *bestMatch()* function takes a list of candidate key frames from database and applies *similarity()* to get the best matched frame. RVF (relation valued function) and the *PerTuple* input mode are defined in [2].

```
@UDF (functionType = RVF, inputMode = PerTuple)
public long bestMatch(KeyFrame sample, List<KeyFrame> keyFrameDate-
Set) {
    long id = 0;
    double bestMatch = 0.0;
    for (KeyFrame f: keyFrameDateSet) {

        double s = similarity(sampleFrame, f);

        if (s > bestMatch) {
            bestMatch = s;
            id = f.getId();
        }
    }
    return id;
}
```

```
@UDF (functionType = Scalar)
double similarity(KeyFrame sample, KeyFrame frame) {
//calculate the similarity between sample and frame
}
```

During the compile phase, functions *bestMatch()* and *similarity()* will be extracted and compiled into individual class files by our annotation processing tool. Code for stub functions will be generated in the original file. During runtime, the individual class files are automatically deployed to specified database as UDFs and are launched on demand (Section 4.4.2). The stubs work as delegates in the flow control main program, communicating with the UDFs. The SQL statement for launching *bestMatch()* can be generated automatically according to the functionType.

Parallelism both at the workflow control level and within each UDF function utilizing multi-core and GPU are beyond the scope of this paper.

4.3.3 Resources Injection

We bind runtime resources (*e.g.*, databases, global index) through the widely applied resource injection technique:

```
@Resource
GlobalIndex globalIndex;
@Resource
DBURI dbURI;
```

The instances containing the runtime information about URLs of database and global index are generated and bound to the variables by our runtime annotation processing tool. The developer only needs to use the resource at the abstract level without knowing any details of the instance implementation.

4.4 Runtime

4.4.1 User Function Container

To let the primitive provider only focus on video process logic, we offer a user function container mechanism to decouple user function development from the DBMS implementation details. The technique has been explored in [2], and we include a brief description here just for completeness. We first introduce RVFs (relation valued functions) into the database, which are a special type of UDFs that accept relations as input. A RVF container isolates the DBMS implementation from UDF developers, by accomplishing relation-object mapping between the database internals and the user functions, and offering the RVF shell, which is a set of APIs dealing with database internal data structures and function calls. The developer only needs to write application logic in user-functions, which can be deployed and run in the RVF container.

4.4.2 Automatic Deployment of UDF

DB functions, extracted from user code by annotation processing tool or provided directly by developers, are stored in the registry (shown in Figure 1). The runtime will run the code or query, issuing sub-queries to individual nodes as necessary, based on the workflow. On each database node, a daemon is deployed to receive such sub-queries. When a sub query includes any db functions not yet deployed locally, the daemon will retrieve the corresponding binary code and deploy it into local DBMS.

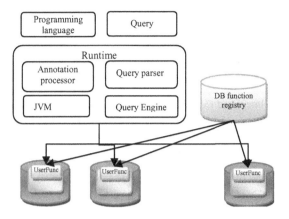

Fig. 1. Multi-database video analytics platform

4.4.2 Automatic Deployment of UDF

We have implemented the entire mechanism into a video analytics platform (Figure 1), the effectiveness of which has been demonstrated through various tests. Figure 2 shows results collected from a typical experiment we ran, where a nearest neighbor search query is executed in a few PostgreSQL databases, as the RVF called in *Block* or *PerTuple/Block* input modes in different runs. Depicted in Figure 2 is the comparison between execution times at an arbitrary database node for the two modes.

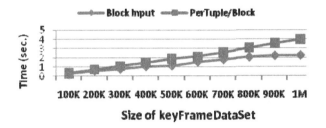

Fig. 2. Execution time of *bestMatch()* at a typical database node

5 Conclusions

Leveraging multi-databases as the computation platform, we have designed a unified programming style, and a database-agnostic runtime. Domain experts can easily and efficiently compose workflow and program domain specific processing primitives using a single programming language, and have the processing primitives dynamically deployed to heterogeneous databases at run time. We believe the proposed methodology will benefit many data-intensive scientific applications where databases are extensively employed.

References

1. http://incubator.apache.org/uima/
2. Chen, Q., Hsu, M., Liu, R., Wang, W.: Scaling and Speeding Video Processing. Submitted to DEXA 2009 (2009)
3. DeWitt, D.J., Paulson, E., Robinson, E., Naughton, J., Royalty, J., Shankar, S., Krioukov, A.: Clustera: An Integrated Computation And Data Management System. In: VLDB 2008 (2008)
4. Gray, J., Liu, D.T., Nieto-Santisteban, M.A., Szalay, A.S., Heber, G., DeWitt, D.: Scientific Data Management in the Coming Decade. SIGMOD Record 34(4) (2005)
5. Nieto-Santisteban, M.A., Szalay, A.S., Thakar, A.R., O'Mullane, W.J., Gray, J.: When Database Systems Meet the Grid. Microsoft Technical Report (2004)
6. Deelman, E., Chervenak, A.: Data Management Challenges of Data-Intensive Scientific Workflows. In: WSES 2008 (2008)
7. Shankar, S., DeWitt, D.J.: Data driven workflow planning in cluster management systems. In: HPDC 2007 (2007)
8. Hadoop, http://hadoop.apache.org/
9. Chaiken, R., Jenkins, B., Larson, P.-Å, Ramsey, B., Shakib, D., Weaver, S., Zhou, J.: SCOPE: Easy and Efficient Parallel Processing of Massive Data Sets. In: VLDB 2008 (2008)
10. Olston, C., Reed, B., Srivastava, U., Kuamr, R., Tomkins, A.: Pig Latin: A Not-So-Foreign Language for Data Processing. In: SIGMOD 2008 (2008)
11. Pike, R., Dorward, S., Griesemer, R., Quinlan, S.: Interpreting the Data: Parallel Analysis with Sawzall. Sci. Program. 13(4), 277–298 (2005)
12. Haas, L.M., Lin, E.T., Roth, M.A.: Data Integration Through Database Federation. IBM System Journals 41(4) (2002)
13. Lowe, D.G.: Object recognition from local scale-invariant features. In: ICCV 1999 (1999)
14. Bay, H., Ess, A., Tuytelaars, T., Van Gool, L.: SURF: Speeded Up Robust Features. Computer Vision and Image Understanding 110(3), 346–359 (2008)
15. Dalal, N., Triggs, B.: Histograms of Oriented Gradients for Human Detection. In: CVPR 2005 (2005)
16. Muja, M., Lowe, D.G.: Fast Approximate Nearest Neighbors with Automatic Algorithm Configuration. In: VISAPP 2009 (2009)
17. A Metadata Facility for the Java Programming Language, http://www.jcp.org/en/jsr/detail?id=175

A High Performance Computing Platform Based on a Fibre-Channel Token-Routing Switch-Network[*]

Xiaohui Zeng[1,2], Wenlang Luo[1], Cunjuan Ou Yang[1], Manhua Li[1], and Jichang Kang[2]

[1] Department of Computer Science and Technology
Jinggangshan University, China
[2] Department of Computer Science and Engineering
Northwestern Polytechnical University, China
zeng_xhui@163.com

Abstract. A high cost-effective Fibre-Channel Token-Routing Switch-Network is designed and developed for the high performance computing fields, and a low-level communication library based on our user-level communication protocol FC-VIA, is appended into the most commonly used MPICH. Thus FC-VIA-MPI (a new extended MPICH) comes into being, and the experiment results show that the FC-VIA-MPI running on Fibre-Channel Token-Routing Switch-Network has achieved better performance than the ScaMPI running on the SCI network.

Keywords: High Performance Computing, Fibre Channel, Virtual Interface Architechture, Message Passing Interface, Scalable Coherent Interface.

1 Introduction

According to our invention patent (patent number: ZL 99 1 15908.x), a Fibre-Channel Token-Routing Switch-Network (abbreviated to FC-net) is designed and developed, adopting the special communication mechanism of our invention patent [1]. The cost of the SCI (Scalable Coherent Interface) NIC (abbreviated to NIC) is about ten times that of our FC-net NIC (the price of SCI NIC is about U.S. $1000 in the year of 2006.).

FC-VIA (FC-net-based Virtual Interface Architecture), a new user-level communication protocol, has been designed and implemented for the FC-net, in which the zero-copy communication mechanism has been implemented and could greatly improve the communication efficiency [2]. MPICH is the most commonly used implementation of the MPI-1 standard [3], and has been ported to many hardware platforms (e.g. IBM, SGI, and Cray) [4].

Because of the special communication mechanism of FC-net, MPICH cannot be used directly for FC-net and the related low-level communication library has to be developed to MPICH. According to the specification of ADI (Abstract Device Interface) of MPICH [5], FC-VIAL (Fiber-Channel-based Virtual Interface Architecture Library) is developed.

---

[*] This work was supported by the National 863 Project of China (No. 2003AA001018).

Y. Wu and Q. Luo (Eds.): ICHCC-ICTMF 2009, CCIS 66, pp. 83–87, 2010.
© Springer-Verlag Berlin Heidelberg 2010

2 Introduction to MPICH

There are three layers in the MPICH. The MPICH's APIs (Application Programming Interfaces) are on the top layer, which provide peer-peer communication and collective communication. In its middle layer, the ADI offers the communication interfaces independent to the physical devices for the MPI applications. In the bottom layer, there are various lower communication libraries [6].

The key of the portability and performance for the MPICH is the ADI which defines a set of library functions to abstract unified communication device [5]; therefore the various complex lower communication devices can be simplified by flexibly invoking appropriate library. The ADI-1 provides four groups of functions to accomplish all kinds of functions. Although ADI-1 increases the portability of MPICH, it also brings some unnecessary software overheads. Hence, the ADI-2 puts emphasis on the performance improvement, redefining some interfaces' syntax, such as contiguous-message operation, noncontiguous-message operation, message polling, etc.

A lower communication library is needed to support the ADI to get a high performance MPI platform, and the overall communication efficiency will be increased prominently by improving the network interface performance and providing user-level access,.

3 Implementation

Depending on the length of the message, the ADI-2 in the MPICH defines three different protocols (named short, eager and rendez-vous) which should be implemented in the lower communication library [5, 7, 8]. This allows designer to make an optimized trade-off between performance and resource usage. The FC-VIAL is the implementation of the ADI-2 in the MPICH for FC-net.

3.1 Architecture of FC-VIA-MPI

To support our high-speed network, a user-level communication protocol FC-VIA is developed, which provides user-level access [2]. To make FC-VIA support the MPICH, the low-level communication library FC-VIAL is designed and implemented; thereby the MPICH is ported to the FC-net. A new MPICH version, FC-VIA-MPI, comes into being. The figure 1 shows the architecture of the FC-VIA-MPI.

3.2 Design and Implementation

3.2.1 Short Protocol
The short protocol is adopted for the messages shorter than certain length. In this protocol, in spite of the permission of the receiver, the users data will be encapsulated into a control packet together with the corresponding message envelop and sent immediately.

According to the FC-2 frame format [9], the frame length can be up to 2148 bytes. In the FC-net NIC, the FIFO (sending and receiving buffer) length is 4096 bytes respectively, so the message no longer than 4096 bytes can be sent in one time by the FC-net NIC. In the FC-VIAL, if the message is no longer than 2148 bytes, the short protocol is utilized to transmit the message.

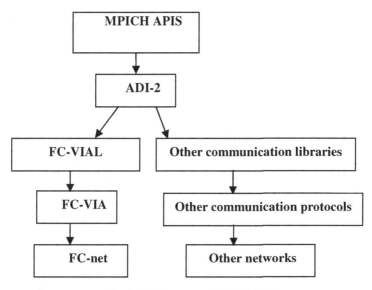

Fig. 1. Architecture of FC-VIA-MPI

3.2.2 Eager Protocol

In our FC-VIAL, the eager protocol is adopted for the long messages between 2148 bytes and 4096 bytes. In that case, needless waiting for the receiver's ACK, the sender will encapsulate the message into a control packet and send the message to the receiving device FIFO.

3.2.3 RENDEZVOUS Protocol

The RENDEZVOUS protocol is utilized for the messages longer than 4096 bytes, in which the data are not delivered until the receiver requests it. In that case, the receiver needn't buffer to store the data, avoiding a lot of data copy. In order to improve communication efficiency, the sender NIC starts sending from the sending FIFO before the sending FIFO has been filled fully. Moreover, the receiver NIC starts reading from the receiving FIFO before the receiving FIFO has been filled fully via write-read-interleave.

The DMA (Direct Memory Address) and memory address mapping technology are adopted to avoid copy (i.e. zero-copy technology) between the Linux user space and kernel space for the three protocols. Hence data can be sent through NIC from user space or be transmitted through NIC to user space without storing data in the kernel space temporarily.

4 Performance

To test the bandwidth and latency, two PC machines, each with an Intel Pentium 2.4G Hz CPU and 512M memory, are installed the Red Hat Linux 9.0 operating system. A Ping-Pong benchmark is used between two MPI processes, each executing blocking send and receive operations to wait for an incoming message (MPI_Recv()) and immediately responding (MPI_Send()) once it has arrived. We have tested for 100 times, and the bandwidth and latency are the mean values. To test the performance of FC-VIA-MPI, the two machines are connected by our FC-net NIC (with 1000Mb/s theoretical bandwidth or so), and the two MPI processes communicate through FC-VIA-MPI. As for ScaMPI (developed by Scali Co.) [10], we use PCI-SCI NIC (made by Dolphin Co. in Norway with 1000Mb/s theoretical bandwidth) to connect the two machines, and the two MPI processes communicate through ScaMPI. The figure 2 shows the bandwidth between the two machines for the FC-VIA-MPI and the ScaMPI, and the figure 3 shows the communication latency.

From the figure 2 and figure 3, a conclusion can be drawn that the bandwidth the ScaMPI achieves is lower than FC-VIA-MPI's, and also the minimal latency is higher for messages transferred via ScaMPI. There are three reasons:

a) The three protocols in the FC-VIA-MPI support different length message; thus the time of packing and unpacking message could be reduced significantly;

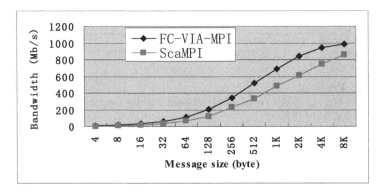

Fig. 2. Communication Bandwidth between Two Machines

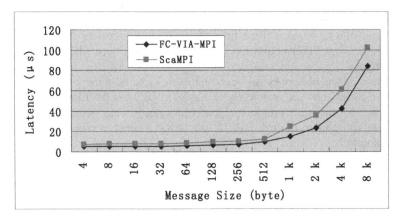

Fig. 3. Communication Latency between Two Machines

b) The FC-VIA-MPI adopts the write-read interleaving technology, which can reduce the synchronization overhead and lead to higher point-to-point bandwidth;

c) The FC-VIA-MPI uses zero-copy technology which could significantly decrease the process switching overhead between the user space and the kernel space.

5 Conclusion

Based on the hardware characteristics of the FC-net, a low-level communication library FC–VIAL has been designed and implemented; thereby the MPICH have been successfully ported to the FC-net, and we gain a new extended MPICH (i.e. the FC–VIA-MPI) and a new high performance computing platform. According to the experiment results, the FC–VIA-MPI can achieve higher performance than the ScaMPI.

References

1. Lei, Y., Zheng, K., Feng, P., Zeng, X., Kang, J.: FC-TRSN: a New Cluster-oriented High-speed Communication Network. In: The 10th WSEAS International Conference on Communications, Athens, Greece (July 2006)
2. Xu, X.: Research on the Communication Protocols of the Cluster Inter-connection Based on the Software/hardware Co-design. [Ph.D. Thesis]. Northwestern Polytechnical University, Xi'an, China (2004) (in Chinese with English abstract)
3. MPICH-A Portable Implementation of MPI,
   http://www-unix.mcs.anl.gov/mpi/mpich/
4. Hwang, K., Xu, Z.: Scalable Parallel Computing Technology, Architecture, Programming. McGraw-Hill Press, America (1998)
5. Gropp, W., Lusk, E.: MPICH ADI Implementation Reference Manual (August 1996)
6. Worringen, J., Bemmerl, T.: MPICH for SCI-connected Clusters. In: Proceedings of SCI Europe 1999, Toulouse, France (September 1999)
7. Gropp, W., Lusk, E.: The Second-Generation ADI for the MPICH Implementation of MPI,
   ftp://ftp.mcs.anl.gov/pub/mpi/workingnote
8. Gropp, W., Lusk, E.: The implementation of the second generation MPICH ADI,
   http://www-unix.mcs.anl.gov/mpi/mpich/workingnote/adi2impl/
   note.html
9. Fibre Channel Framing and Signaling,
   ftp://ftp.t11.org/t11/pub/fc/fs/03-173v1.pdf
10. Scali MPI – ScaMPI, http://www.dolphinics.com/pdf/documentation/

XML Application for SNMP_based Information Management

Wenli Dong

Institute of Software, Chinese Academy of Science
Beijing, China
Wenli0306@yahoo.cn

Abstract. The information management requires flexible, extensible form as information continues to grow in size and complexity. This paper analyzes the architecture of the XML application for SNMP_based information management and study how to combining XML with SNMP_based information management. Simple HTML requirement is sent to web server, the particular management interface are provided after XML/MIB mapping, XML files of the response are created to display on the HTML page based on XML technologies.

Keywords: Information Management, Simple Network Management Protocol, eXtensible Markup Language, Management Information Base.

1 Introduction

The information management requires flexible, extensible form as information continues to grow in size and complexity. The information systems developed and maintained by different operators are implemented by different techniques, and they are deployed and supported by different platforms. While the Simple Network Management Protocol (SNMP)[1][2] is still the dominant information management technology, SNMP_based information management is limited by its information form. XML (eXtensible Markup Language) is an ideal solution for resolving above problems. By applying XML technology in SNMP_based information management, adopting XML in describing information could add /delete/modify MIB (Management Information Base) dynamically and flexibly. Furthermore, the technologies of XML//Xpath/ XSL (Extensible Style Language) can provide user a structuralized, flexible, and accurate means to specify and query information.

Many researchers study the how to use XML in SNMP_based information management [3][4][5][6][7][8][9], but most of these efforts concentrate on information transformation between XML and MIB. They are so simple that they cannot give enough information to understand the XML application for SNMP_based information management and can not provide a strong basis for the information management architecture. This paper analyzes the architecture of the XML application for SNMP_based information management and study how to combining XML with SNMP_based information management.

The rest of the paper is organized as follows. Section 2 presents the architecture of the XML application for SNMP_based information management. Section 3 introduces

Y. Wu and Q. Luo (Eds.): ICHCC-ICTMF 2009, CCIS 66, pp. 88–94, 2010.
© Springer-Verlag Berlin Heidelberg 2010

the XML and SNMP gateway. Based on the parsing, the mapping method is presented. In the subsequent two sections, some related knowledge is presented, that is, SNMP_based information management and Web server and Servlet (Server Applet). Section 6 summarizes the paper.

2   Architecture of the XML Application for SNMP_based Information Management

Figure 1 presents an outline of the system structure that can be used to manipulate information [10][11][12][13]. It consists of four parts: The data layer, database management system (DBSM), is used to store information. Logic layer, XML/SNMP gateway, is used to accomplish the information exchange between XML_based information and MIB_based information. Web layer, Web server, provides the services, and the web component in the server can accomplish some work directly. Client layer, XML_based web browser, provides a user-friendly way for user to manipulate the information. The web browser are designed to be accessed by a user through graphical interfaces, it is a web interface for the user to send command for adding, deleting, querying the information stored in database, and the services are listed on the HTML pages. The client code is used to support the operation sent by Web browser.

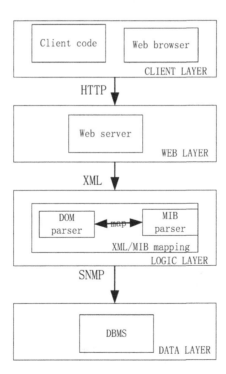

Fig. 1. System Structure

3 XML and SNMP Gateway

XML is self-describe, structuralized language. It is flexible, extensible, and applicable to different applications. The Schemas may be used to describe a complete information structure. Besides the XML Schemas, a set of precise translation rules that can be used to produce a simulation script for displaying and querying are defined. These translation rules have been expressed by means of the XSLT(eXtensible Stylesheet Language Transformation) transformation language and have been implemented by a JAVA servlet. The XSLT transformation document has been organized in independent files, according to the structure of the Schemas presented. Xpath provides a flexible way of addressing (pointing to) different parts of a document.

SNMP is a widespread acceptance network management protocol existing to manage a large number of network devices. So, the demand for resolving the problem of interoperation between XML_based information and MIB_based information. The XML/SNMP gate way is designed and developed to implement the information exchange between XML manager and SNMP agent.

SNMP agent visits the managed object by traversing nodes in MIB tree [1][2]. A Management Information Base (MIB) is a collection of information that is organized hierarchically. MIBs are accessed by using a network management protocol such as SNMP. They are comprised of managed objects and are identified by object identifiers.

A managed object (sometimes called a MIB object, an object, or a MIB) is one of any number of specific characteristics of a managed device. Managed objects are comprised of one or more object instances, which are essentially variables. MIB defines the variables maintained by SNMP agent. The variables are read and written to implement the network management.

XML/SNMP gateway is used to accomplish the information exchange between XML_based information and MIB_based information [14]. As has been discussed above, SNMP accesses managed object by traversing MIB tree. The main task of XML/SNMP gateway is the mapping between XML Schema and MIB.

3.1 XML Parser

XML parser and MIB parser are necessary for the mapping.

There are two primary ways that applications can pull information from a XML parser:

DOM parsers [15] parse the XML document and create an object-oriented hierarchal representation of the document, which can be used at run-time. The structure is easy to use but the technique is very resource demanding both in terms of memory and CPU.

SAX parsers [16] do not store any information. Instead they scan the information and call handler functions that are associated with specific instructions and tags in the XML-document. The SAX parser is often faster than the DOM parser and is not as resource intensive. The drawback is that the logic for handling the XML document instructions and tags is significantly more complex.

Because SAX can't load files in memory, so it can't access the files at random, but DOM can. Usually DOM can: 1) Being familiar with the Schema structure; 2) Moving the part of Schema (such as arranging in an order); 3) Using the information in Schema more than once. Usually SAX can: 1) Collecting some elements only from one Schema file; 2) Not taking up too much memory; 3) Using once for the information of Schema's.

Because while mapping between XML and MIB, the parsed information will be visited more than once to get the information in memory, such as element, attribute etc. to be assembled as MIB. DOM way, using DOM parser to pull information from a Schema is employed for the mapping. After been parsed by DOM parser, a Schema is turned into a memory tree called DOM tree. The MIB parser visits this tree, and gets the relevant element, attribute information, then turns into a MIB tree.

3.2 MIB Parser

The MIB parser maintains its MIB as a tree of JAVA objects that reflects the structure of the OID (object identifier) tree. In the MIB tree, leaf nodes represent items of management information that cannot be further subdivided (columnar or scalar objects in SNMP parlance). All other objects are internal node. The objects in a MIB actually represent classes of management information that may have one or more instances. The logic layer stores the code to perform a GET or SET on an instance of an object in the node for that object. Any additional information concerning an instance of an object is stored in a cell of an array indexed by the instance number, and the array is stored at the object node. When logic layer receives a GET or SET request, it takes the OID argument to the request and traverses the MIB tree until it finds the node having that OID. Because only instance of scalar or columnar objects can be the targets of GETs or SETs, this node must be a leaf node. Logic layer then retrieves the code for performing the GET or SET and executes it, passing in the instance number as a parameter. The GET or SET code can use the instance number to access the array to retrieve information associated with the instance, which could even be more code to execute, specific to that instance. A new instance for an object can be created by doing a SET on the object with the instance number of the new instance.

3.3 The Mapping between XML_based Information and MIB_based Information

An MIB tree is generated at startup from an XML document describing the MIB. The XML parser parses the document to create a representation in JAVA object of the DOM tree, extract the information from DOM tree to create a MIB tree. On the other hand, the MIB parser parses MIB, extracting the information from MIB tree to create a DOM tree and then form XML document.

Once defined the structure of a XML document through the Schema file, the next step is the creation of a file with the information in the XML format. This information is collected by the agents created for this task. With the definition of the information described in the Schema file and with the information stored in the MIB, the XML file will be generated automatically.

4 SNMP_based Information Management

New MIB objects are added to the tree through the code in logic code, the SNMP PDU which performs the SET operations contains the OID for each proxy followed by a string containing the value to which it is set.

The value to which the logic layer SET is a string containing a description in XML of the new objects in the sub-tree. This staring is parsed in the same manner as the XML document describing the MIB at startup, and a sub-tree of object is produced. This new sub-tree is then grafted onto the MIB tree.

Deleting of MIB has the effect of removing the entire sub-tree rooted at that node. The XML string passed as a value contains the simple command "delete" and the OID of the node to be deleted.

5 Web Server and Servlet

The web server consists of a servlet running on a web server. Servlet are server-side analogs to Java applets, and allow new server functionality to be added seamlessly (without interrupting web server operation) at run time. Servlets receive the http requests that are addressed to them from the web server, and then do some servlet-specific processing on them.

The user enters the managed element's address in an HTML form on a web browser and submits it to a web server running the servlets. The web server passes the form to the servlet, which then contacts the DBMS by sending an SNMP GET request .If the web server doesn't receive a reply within a timeout period, it sends a web page back to the user's browser explaining that it is unable to reach the agent. If it does receive a replay, the web server extracts the reply's payload which contains a XML sting describing the MIB, then generates an object representation of the XML_MIB.

The user requests the addition or deletion of variables in an MIB through HTML forms generated by the web server and displayed on the users browsers .The web server then sends to the logic layer to carry out these requests. In response to each set, the logic layer returns an XML string describing the state of its MIB after these are performed. This XML string is, in a sense, the "value" of the MIB variable, and it is consistent with SNMP semantics to return the new value of a variable whenever a set is performed on that variable. The logic layer use these XML string to update the user's graphical display of the MIB, and also to generate new HTML forms for the user's browser which provides choices to the user as to which variables can be uploaded to the agent, deleted from the DBMS, or monitored.

6 Conclusion

This paper studies a solution based on XML and SNMP for a system open and multi-platform that allowed the management of heterogeneous systems dynamically. XML introduces great potential for system management; being highlighted the transparency between consultation and the implementation of the information made available. In this situation, the user or application does not need to know how the information is stored internally. The introduced solution offers compatibility with the management systems already existing through automated tools, a MIB already defined and Standardized was quickly converted to XML documents with the minimum effort.

This paper analyzes the architecture of the XML application for SNMP_based information management. The Browser_based user-friendly interface brings a new layer of clarity and efficiency to information management. The XML/SNMP gateway is designed and developed to accomplish the information exchange between XML_based information and MIB_based information. Simple HTML requirement is sent to web server, the particular management interface is provided after XML/MIB mapping, XML files of the response are created to display on the HTML page based on XML technologies.

Acknowledgment

This work is supported by the National High Technology Research and Development Program of China (Grant No.2007AA01Z190).

References

1. IETF RFC 1157, Simple Network Management Protocol, SNMP (1990)
2. IETF RFC 2578, Structure of Management Information Version 2, SMIv2 (1999)
3. Changqing, Z., Qinming, H.: The Reserch and Implementation of SOAP/XML_based Network Management System. Computer Applications and Software 9, 96–98 (2005)
4. Wang, Z.-h.: Use VB and Java Web Service Carries on the Internal Network Management. Computer Knowledge and Technology 17, 29–31 (2006)
5. Lei, X., Yue, J.-h.: Research on Network Management Implementation Mode Based on Web Services. Computer and Modernization 3, 45–48 (2006)
6. Feng, X., Song, R.-s., Zhao, Q.-s.: Study and Implementation of Mixed Network Management Based on Web Service. Application Research of Computers 7, 125–127 (2004)
7. Gong, X.-h., Xiong, Q.-b.: Web Service-based Network Management. Computer Application 10, 78–81 (2003)
8. Peng, Y.: Telcommunication Network Management Based on Web Services. Data Communication 4, 9–11 (2003)
9. Yuan, Y., Guoxing, J.: The Research of Network Management with Web Service. Journal of Taiyuan Teachers College 3, 29–3 (2005)
10. Dong, W.: Dynamic Reconfiguration Method for Web Service Based on Policy. In: International Symposium on Electronic Commerce and Security, pp. 61–65. IEEE Press, New York (2008)
11. Dong, W., Jiao, L.: QoS-Aware Web Service Composition Based on SLA. In: The 4th International Conference on Natural Computation, pp. 247–251. IEEE Press, New York (2008)
12. Dong, W.: SLA Monitoring Based on Semantic Web. In: Qi, L. (ed.) Applied Computing, Computer Science, and Advanced Communication. LNCS, vol. 34, pp. 21–29. Springer, Heidelberg (2009)

13. Dong, W.: Construction and Test of Web Service Solution for E-government. In: International Conference on Computational Intelligence and Natural Computing, pp. 221–224. IEEE Press, New York (2009)
14. Dong, W.: Translation from GDMO/ASN.1 to tML/Schema. Journal of Convergence Information Technology 3, 17–25 (2008)
15. W3C Document Object Model (DOM)
16. Megginson David SAX 2.0:The Simple API for XML

A Method of Synthetic Scheduling Basing on the Grid Service

Hongwei Zhao and Xiaojun Zhang

School of Information Engineering, ShengYang University, ShenYang Liaoning, China
zhw76930@sina.com

Abstract. In order to implement the balanced distribution of the Service Grid system and to improve the utilization ratio of the resource as well as handling up rate of the system, a method of Synthetic Scheduling basing on the Grid service has been designed and implemented after the study on the Grid service. Firstly, a layered loading balancing scheduling mode has been proposed, providing the structure of the dispatching service system. Secondly, a comprehensive service distribution algorithm has been designed and implemented in consideration of respective local service counts, each join points' performance and current load distribution. Finally, the result of the experiment indicates that the scheduling system can improve the efficiency of dispatching service and the utilization ratio of distributed grid resource.

Keywords: Distribution; Service Grid; Synthetic Scheduling; Load balance.

1 Introduction

Grid is a type of integrated resource and service environment, which includes various related resource and service integrated with computing power, data information and knowledge, software and people, aiming at organizing the geographically spreading computers into a "virtual supercomputer" with the use of internet. Service Grid is a special type of grid, which improves the application of the Web Service technology, provides a range of resource sharing to the extreme for enterprises and is on the way to become the most effective way to increase the overall level and capacity of enterprise.

At present, there exists a great deal of research on grid in all aspects, ie., relatively well-known systems, such as Globus, Legion, Condor, and so on, with the developing aim to take the effective use of the geographical distributed resource, in which effective scheduling strategy seems very critical for optimizing resource utilization rate.

Similarly, in the process of service schedule of service grid, we also need to coordinate the use of the distributed resource to carry out transparent and automatic service adjustment by a number of local agents. In order to implement the balanced distribution [1] of the Service Grid system and to improve the utilization ratio of the resource as well as handling up rate of the system, how to realize the service distribution has become the core of the mechanism of grid system, The distribution of service depends on the access to the load information of calculation nodes and processing technology. Therefore, how to access the load information of nodes, how to measure

Y. Wu and Q. Luo (Eds.): ICHCC-ICTMF 2009, CCIS 66, pp. 95–100, 2010.

and evaluate the load condition of local agents with the use of above mentioned information for service distribution have become the important load-balancing system research.

2 Related Work

In Service Grid [2,3] system, service is dynamically generated with the size of the load of each local agent changing dynamically, thus only dynamic load balancing scheduling strategy is usually put into consideration instead of the static load-balancing scheduling method. In general, dynamic load balancing scheduling can be divided into two broad categories, centralized scheduling and distributed scheduling [4], All of the service of the first one is submitted to the global agent, which will be in charge of collecting load information of the relative local agent to determine the load-balancing scheduling program. In this mode, local agent does not carry out scheduling but to distribute the service assigned by the global agent and submit the service back to the global agent when it ends, the main advantage of which lies in its relatively simple realization and the disadvantage of which lies the high costs of scheduling for the global agent would become the bottleneck of the system in the environment of large–scale service grid with a great number of nodes. Each of the local agent of the second one can receive service and carry out scheduling, realizing load balancing operation according to some of the load information in the local scope, by which each computer would broadcast its load information to others on a regular basis to update those of the local maintenance Load vectors, the biggest advantage of which is good scalability and the drawback of which is the long time to wait for service as a result of the large amount of communication among nodes. Aiming at the characteristics of the change of service grid load [5], this paper first proposes the layered load balancing scheduling model on the basis of the analysis of the load-balancing scheduling model [6,7], then brings about the structure of this system, and at last designs and achieves a type of service distribution algorithm which comprehensively taking the number and the performance of relative local agent service as well as the current load situation into account.

3 Layered Scheduling System Architecture and the Realization of the Algorithm

3.1 Layered Scheduling System Architecture

According to the pros and cons of the above mentioned two kinds of scheduling modes, we have brought out a layered scheduling model, in which the global agent is in charge of collecting load information of the relative local agent and all of the service is submitted to the global agent, but different from centralized scheduling, not all of these tasks are saved in the global agent service submitted queue waiting for scheduling, but are directly assigned to local agents by global agent in accordance with load balancing and scheduled by local agents. Thus the global agent will not interfere with the service and its load reduce , which avoids becoming the bottleneck in the system with its less service waiting time, in order to achieve a simpler realization than distributed scheduling. From the view of the whole service grid system, taking centralized scheduling in local parts and the distributed scheduling in global ones would not only maintain the advantages of centralized scheduling, but also make up for the deficiencies of it in the use of distributed scheduling on the overall situation layered scheduling system structure is composed of the following parts:

1) Service Composition Module, receiving service requests and achieving a reasonable dynamic service composition according to the status of each node portfolio of services.

2) Scheduling module, a receiving module to a distributor, in charge of dynamic collection of load information on various nodes, setting up the distribution levels of nodes and transmit the information to service distribution module on a regular basis by the analysis on the performance of node, node CPU utilization, memory usage and I / O usage, and so on.

3) Monitoring module, monitoring whether the local agent overload or delay too long to start re-scheduling strategy.

4) Transmission module, transmitting information of each node and integrated performance level of service by the use of HTTP transmission based on grid technology, since the majority of Internet firewall and proxy will not undermine the HTTP transmission.

5)Database module, requiring constant update because of the changing parameters such as the various indicators and comprehensive performance of local agent, and the number of service dealt by local agent.

3.2   Load Balancing Principle of Layered Scheduling System

The main principle of Layered load balancing scheduling is that the global agent would receive all requests for service, and then distribute them to local agents to implement scheduling based on certain principles, the main purpose of which is to enable the local agent perform a more balanced load distribution in order to obtain a higher overall handling capacity and faster response speed. At present the common methods of request distribution mainly conclude three of them, such as running and turning means, least connection means and fastest connection means. By the first kind of methods, it is simple to achieve but the problem of load balancing has not been put into consideration in essence. By the second one, the performances of various the servers have not been dealt with distinctly. By the third method, the current load condition has not been taken into account though the performance of the server has been considered. So these limitations make these algorithms fail to achieve the load balancing distribution.

In order to improve the efficiency of service distribution algorithm to adapt to service grid system, the global agent should be able to know the processing power of each local agent, and accurately track the load condition of various local agents simultaneously. Based on the above mentioned requirements, a method of synthetic scheduling has been designed in this article which comprehensively taking the number and the performance of relative local agent service as well as the current load situation into account.

3.3 Realization of the Method of Synthetic Scheduling in Layered Scheduling System

In grid environment, there are a great many grid nodes and resource and a much complicated matching relationship between service application and service resource. For example, a user may contain a service request and multiple ones as well; some individual service may be composed of multiple sub-services which having some kind of dependent relationship among them or independent of each other. Diversified service requests submitted by different users may be implemented in the same grid resource, forming the relationship of competition; also, a service processing may need to simultaneously or successively use multiple grid resource. The relationships among services and services, service and grid resource nodes, resource nodes and nodes will eventually affect the synthesis of grid service.

When users submit service requests to grid system, grid resource location and a copy of positioning will first check appropriately to determine what kind of sub-services existing in grid system and what to be re-developed if there is no direct service resource available to grid system while the requested service associated by a number of related services, invoking directly the services existing in grid system and redeveloping those non-existent ones following the requirements of grid system(The specific process of developing grid services is omitted here) Secondly, the user's request is to be described as a process composed of multiple sub-services with the use of pre-defined correlation structure, a sequential structure forming when a service needs the implementation result of a related service as its prerequisite; a cluster structure constituting if a service requires implementation results of more than one service as its prerequisite, and a branch structure forming if the implementation result of a service works as prerequisite of other two or more than two services.(all sub-services being available in this article)

Through the identification of associated structures among all sub-services, match scheduling can be performed according to the service to be implemented. In grid environment, because of the large-scale characteristic of resource and the copy management of resource as well, there may be more than one requested service resource to meet some requirement, which may obtain different abilities and pay different costs during the implementation in the grid resource at the same time, ie., there exists difference in the quality of service offered by the same qualified service resource, which is significant sometimes. Therefore, the optimal match scheduling of service should be taken into account for grid synthesis service scheduling. After match scheduling done, service can be performed, whose implementation process being under the control of the local service management system. Following the implementation process, the occupied resource should be returned to internet resource management part, while the grid Service management module deliver performance results and related information to the submitter or directly to the next service node to be implemented the, with the use of RSL language to describe the resource information and related parameters to transmit information.

In summary, this paper presents the strategy of grid synthetic service scheduling as the following:

To determine the sub-services included in the service submitted by users;

To determine what kinds of sub-services already existing, what to redevelop according to grid demand and how to develop.

To determine the associated relationship among the various sub-services, with scheduling process including:

a. In accordance with the implementation of the different services from the execution services to be implemented, adjusting the order of service scheduling to make different services matching carried out at the same time, thereby reducing the processing time of grid's request on users, known as order schedule.

b. finding the appropriate service resource for the services to be performed (ie, select the copy resource), known as the match schedule.

c. implementing the matched services, followed by transmitting scheduling request and data information to the service to be scheduled.

4 Experiment and the Analysis of Results

In this paper, the project kit, GridSimtoolkit4.0, has been used in the simulation experiment, mainly because GridSim[8] acting on the simulation test focusing on the scheduling strategy in the grid environment, providing the various basic function components of grid and simulating the various basic actions of the function components, which making the developers achieve scheduling simulation with ease by this simulation tool. And related service grid association has been simulated. In the simulation experiment, response time has been compared between running and turning means of the service distribution algorithm and that mentioned in this paper considering load balancing, the two curves are as follows, (shown as Fig 1)

 a: the theoretical value

 b: means of service synthetic scheduling presented in this paper

The Experiment shows that the synthetic scheduling method proposed in this paper, ie., services being dynamically allocated, can enable all local agents load balanced, accelerate the speed of service scheduling shorten efficiently the completion time of service, and reduce the impact of service delay on the improving the parallel efficiency of

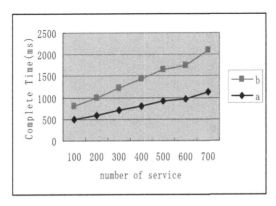

Fig. 1. The comparison chart of the theory and practical without delay overall service grid

5   Conclusion

In this paper, a method of Synthetic Scheduling supporting service grid is proposed in this paper, followed by the analysis of the specific model and technology related to service grid, together with a layered scheduling model and the structure of load-balancing system, and then a method of Synthetic Scheduling comprehensively considering the task number and current load performance of various local agents has been designed and realized. The main purpose of this method is to improve the efficiency of the implementation of grid system, by which to solve the problems of service grid system such as scheduling delays and so on. Finally, the superiority of the method proposed in this paper to other methods is verified by the experiment.

References

[1] Andrew, J., Naughton, T.J.: Dynamic task scheduling using genetic algorithms for heterogeneous distributed computing. In: Proceedings of the 9th International Workshop on Nature Inspired Distributed Computing, IPDPS, April 2005, pp. 189–197 (2005)

[2] Mealor, P.D.: Design document for Grid Services Performance Measurement Point, May 13 (2003)

[3] Foster, I., Kesselman, C., Tuecke, S.: The anatomy of the grid Enabling scalable virtual organizations. International Journal of Supercomputer Applications 15(3), 200–222 (2001)

[4] Maheswaran, M., Siegel, H.J.: A Dynamic Matching and Scheduling Algorithm for Heterogeneous Computing Systems. In: Seventh Heterogeneous Computing Workshop, pp. 57–69. IEEE Computer Society Press, Los Alamitos (1998)

[5] He, Y.X., Liu, Z.Y., Deng, A.L.: Load distribution and balancing strategy of distributed system based on workstations. Computer Engineering 25(11), 11–13 (1999)

[6] Li, D.M., Shi, H.H., Gu, L.Q.: Layered load balancing scheduling model based on Rules. Computer Science 30(10), 16–20 (2003)

[7] Gu, L.M.: Research on Cluster Server load balancing technology. Micro Computer Information 23(12), 20–23 (2007)

[8] Buyya, R., Manzur, M.: GridSim: A Toolkit for the Modeling and Simulation of Distributed Resource Management and Scheduling for Grid Computing. The Journal of Concurrency and Computation: Practice and Experience 14(13-15), 1175–1220 (2002)

A Data Hiding Method for Text Documents Using Multiple-Base Encoding

Chin-Chen Chang[1,2], Chia-Chi Wu[1], and Iuon-Chang Lin[3]

[1] Department of Computer Science and Information Engineering,
National Chung Cheng University, Chiayi 621, Taiwan, R.O.C.
[2] Department of Information Engineering and Computer Science,
Feng Chia University, Taichung 40724, Taiwan, R.O.C.
[3] Department of Management Information Systems,
National Chung Hsing University, Taichung, Taiwan, R.O.C.
ccc@cs.ccu.edu.tw

Abstract. In this paper, we propose a novel data hiding method for text document using multiple-base encoding. The combination of the repeated words in the cover-text is used to compute multiple-base quotas. Then the secret data will be transformed into the expression of multiple-base expressions. Finally, some locations of the repeated words will be shifted to generate the stego-text. According to the experimental results, our method is flexible for most documents and can slightly modify the inter-word space. Moreover, our method possesses superiority to the others in terms of embedding capacity and imperceptibility in the text data hiding.

Keywords: Information hiding, multiple-base encoding, text document.

1 Introduction

Data hiding is a celebrated research field of computer science, which embeds the secret data into a digital medium with imperceptibility, and then only the authorized receiver can extract the hidden secret data. The digital medium is called the cover-medium, such as a cover image or a cover text. Once the cover-medium has been embedded with the secret data, it is called the stego-medium. The main purpose of data hiding is to make the casual readers or the malicious attackers unsuspicious for the existence of secret data in the stego-medium. Therefore, data hiding techniques usually select the meaningful media as the cover-media and employ the imperceptibility of the human visual system (HVS) [1] to create the stego-media.

Generally speaking, data hiding techniques can be applied to two applications: the intelligent property protection and the confidential communication. For the intelligent property protection, most researches develop the robust digital watermarking techniques to protect the copyright [1], [2], [3], [4], [9], [12] and [13]. For the confidential communication, the researches concern the imperceptibility and hiding capacity to protect the secret data, such as [5], [6], [7], [8] and [12]. In this paper, we study the latter topic.

Y. Wu and Q. Luo (Eds.): ICHCC-ICTMF 2009, CCIS 66, pp. 101–109, 2010.

The hiding capacity is the quantity of secret data that can be hidden in the cover-medium. On the other hand, the imperceptibility represents that the stego-medium is visually indistinguishable from the cover-medium. We expect an ideal data hiding scheme that can possesses the hiding capacity as large as possible, and effects the least distortions in the stego-medium.

Many data hiding schemes [5], [6], [7], [8], [10] have been proposed in recent decade. In these schemes, they usually use three kinds of digital objects, such as image, audio, and video. Especially, the digital image is composed of pixels. Even though some pixel values are slightly modified, the human eyes are hard to percept these tiny changes. Hence, during nowadays, the researches of data hiding are very popular for still images.

Generally speaking, data hiding in still images can be classified into two domains: spatial domain and frequency domain. Most methods in spatial domain are directly to modify the least significant bits (LSBs) 0 of the pixels in the cover image for hiding data. Therefore, the hiding capacity is determined by the size of cover-image and the number of hidden bits. However, while the modification of LSBs is more than 3 bits, the distortion of the stego-image can be found easily.

On the other hand, data hiding in frequency domain first transforms the cover-image into frequency spectrums consisting of a number of coefficients via some transformation techniques such as discrete cosine transformation (DCT) 0 and discrete wavelet transformation (DWT) 0. Then the sender embeds the secret data by changing some frequency coefficients. This advantage is that the embedded data are not easy to be destroyed, but the data hiding capacity is rather little.

Recently, some hiding methods concentrate on the studies of embedding secret data in text document. Since digital images have a large quantity of the redundant information, a tiny modification of these redundant data can be invisible by the human visual system. However, the text data hiding is difficult since the redundant information of a text document is less than other media. In 1996 Bender et al. 0 proposed three kinds of data hiding methods for text documents: open space methods, syntactic methods, and semantic methods. After that, several researches follow these three concepts to improve hiding capacity 0 or propose a variety of applications 0.

On concerning the open space methods for text hiding 0, there are still some drawbacks as shown in Section 2. Therefore, in this paper, we propose a novel text hiding method based on multiple-base encoding. The proposed scheme first counts the frequency of the repeated words in the cover-text, and then employs the multiple-base encoding concept to change the space of the repeated words to generate the stego-text. The experiment shows that our method can be applied to many kinds of documents.

The rest of this paper is organized as follows. In Section 2, we briefly review the related text data hiding methods. The background and drawbacks of these methods will be discussed. In Section 3, we present our proposed scheme. In Section 4, we demonstrate the experimental results. Finally, some conclusions are remarked in Section 5.

2   Related Works

In this section, two famous text data hiding methods will be introduced. Furthermore, some drawbacks on these methods are described.

2.1 Bender et al.'s Data Hiding Method

Bender et al. proposed three kinds of text hiding methods by opening space 0, which includes inter-sentence spacing, end-of-line spaces, and inter-word spacing. We briefly review inter-word spacing method with the most hiding capacity among them.

This method exploits inter-word space of text to encode data. First, the secret data must be transformed into binary form. Then the sender employs the extra spaces between words to embed the secret bits. For example, one space can be interpreted as a "0", two spaces are translated as a "1". The main advantage of this method is to obtain larger hiding capacity. However, the attacker can easily collect the embedded bits and removed these data formats.

2.2 Chen et al.'s Data Hiding Method

In 2006, Chen et al. proposed a text hiding method 0 with stego-encoding using TeX tool. They grouped each pair of consecutive words as a single word, as shown in Fig. 1. Then they embed one bit of encoding data into inter-word space of each grouped pair. We briefly introduce their encoding scheme as follows.

Keep spaces between groups unchanged.

If they want to embed a bit 0, the inter-word space of the grouped pair is unchanged.

If the bit 1 must be embedded, then the inter-word space is widened for the first grouped pair occurrence, and then shrunk and widened alternately for the next 1's. In Fig. 2 0, a bit string, 101111010 is embedded in a text with 24 inter-word spaces. The plus sign indicates widening; a minus sign means shrinking and blank means unchanging.

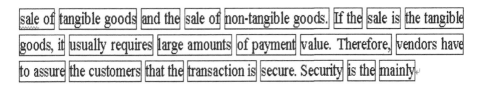

Fig. 1. Grouping of words

	1	0	1	1	1	1	0	1	0
Embedded bits									
Space changes	+		−	+	−	+		−	

Fig. 2. Chen et al.'s embedding scheme 0

Although Chen et al. have taken a consideration to keep the total of space widths between the original text and the embedded text; it is still possible to make some differences in the last pair of each line from the original text. These side effects will result in error of data extraction.

3 The Proposed Scheme

In this section, we propose a text data hiding scheme with high capacity and imperceptibility. This method can improve the drawback of Chen et al.'s scheme and make the least modification of cover-text with hiding capacity. To do that, we first count the number of the repeated words in the cover-text, and then build repeated word sets W_i to record their corresponding positions, respectively. Second, we check whether the cover-text capacity can accommodate the secret data size. After that we perform the embedding data procedure.

We assume that a symmetric key K is pre-shared between the sender and the receiver. The sender encrypts the secret data with the symmetric key K for confidential transmission.

In Section 3.1., we first explain multiple–base notational system. Then we show the embedding phase and extracting phase in Section 3.2 and Section 3.3.

3.1 The Multiple-Based Notational System

All integer numbers can be expressed using multiple-base notational system as follows.

$$x = (d_n d_{n-1}...d_1)_{b_n b_{n-1}...b_1}, \ where \ 1 \le d_i < b_i (i = 1,...,n). \tag{1}$$

These $b_1, b_2,...,b_n$ indicate the different bases corresponding to the symbols $d_1, d_2,..., d_n$. We can compute the decimal value x as follows.

$$x = d_1 + \sum_{i=2}^{n} (d_i \cdot \prod_{j=1}^{i-1} b_j) \tag{2}$$

The decimal value x also can be covered into the multiple-base notational system according to the given $b_1, b_2,...,b_n$ as follows.

$$d_1 = x \bmod b_1 \tag{3}$$

$$d_k = \frac{1}{\prod_{j=1}^{k-1} b_j} \left[x - d_1 - \sum_{i=2}^{k-1} \left(d_i \cdot \prod_{j=1}^{i-1} b_j \right) \right] \bmod b_k, k > 1. \tag{4}$$

Equations (3) and (4) can be used to generate a multiple-base notations with a set of given bases; for example, $48=(1300)_{3532}$ and $157=(3141)_{6254}$.

In our method, we transform the secret massage to a variable base system to find appropriate shifting position in the repeated words of text file. Of course, our method also can apply to TeX tool as the same imperceptivity as Chen et al.'s scheme.

3.2 Embedding Phase

In this section, we propose Algorithm 1 to perform the embedding phase. For clear description, the related data structures will be explained as follows.

We first scan the cover-text (as Fig. 3) to construct a *Text linked list* (as Fig. 4) which comprises four fields: *Serial number*, *Content*, *Shifting bit*, and *Next pointer*. The *Serial number* stores the serial number of the node, and *Shifting bit* default value is zero.

Then we calculate the repeated words from the linked list to generate a *Correlation table*. We collect different repeated words to store a data set $W=\{W_1, W_2,..., W_n\}$, where n is the number of different repeated words and then record their occurrence numbers and serial numbers in the *Correlation table* respectively such as Table 1. $|W_i|=b_{i, 1\leq i \leq n}$ denotes the size of W_i, and W_{ij} means the jth word in W_i.

Of course, we must make sure the cover-text hiding capacity before embedding. Assume that the cover-text has $\sum_1^n b_i$ repeated words, so there are $\prod_{i=1}^n b_i$ shift ways.

Hence, they represent the data hiding payload from 1 to $x = d_1 + \sum_{i=2}^n (d_i \cdot \prod_{j=1}^{i-1} b_j)$.

If the cover-text data hiding payload is smaller than the secret data, then the secret data have to be divided into the proper segments to embed the different cover-texts, respectively. If the payload of the cover-text is enough, we explain the message embedding algorithm as follows.

Algorithm 1. Message Embedding by Multiple-Base Encoding

Input: a cover-text C; an encrypted message whose decimal value is D using the key K.

Output: a stego-text S.

Steps:

1. Scan the cover-text to construct the *Text linked list*.
2. Count the repeated words from *Text linked list* to make a *Correlation table* which contains $W=\{W_1, W_2,..., W_n\}$ and $|W_i|=b_{i, 1\leq i \leq n}$.
3. Compute $x = (d_n d_{n-1}...d_1)_{b_n b_{n-1}...b_1}$ from D by using Equations (3) and (4), where x is a multiple-base notational expression.
4. Find the associated W_{ij} with b_i and d_j, and then record its *Serial number* respectively. Afterward we write "1" into the *Shifting bits* in all recorded *Serial numbers*.
5. Sequentially scan and output a stego-text according to *Text linked list*: if the *Shifting bit* equals 1, then we output two blanks after this word. Otherwise, we output a blank in the latter.

We can investigate that we only can shift 3 sets to represent $x=1+1*4+5*3*4=65$ integer values through this algorithm. Therefore, this scheme is efficient and practical with high hiding capacity. After the value $(65)_{10}=(511)_{534}$ is embedded, we have a stego-text as in Fig. 5. We can find their serial numbers are 3, 4, and 68. For convenient observation, we employ a symbol "^" to represent a blank.

Why are NP-complete problems interesting? First, although no efficient algorithm for an NP-complete problem has ever been found, nobody has ever proven that an efficient algorithm for one cannot exist. In other words, it is unknown whether or not efficient algorithms exist for NP-complete problems. Second, the set of NP-complete problems has the remarkable property that if an efficient algorithm exists for any one of them, then efficient algorithms exist for all of them.

Fig. 3. The cover-text (from *INTRODUCTION TO ALGORITHMS* 0 Page 9)

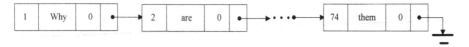

Fig. 4. Text linked list

Table 1. *Correlation table* of the cover-text

i	Word (W)	Occurrence number(j)	Serial number
1	NP-complete	1	3
		2	13
		3	44
		4	50
2	Problems	1	4
		2	45
		3	51
3	Efficient	1	9
		2	25
		3	40
		4	59
		5	68
4	Algorithm	1	10
		2	26
		3	60
5	For	1	11
		2	27
		3	43
		4	62
		5	71
.	.	.	.
.	.	.	.
.	.	.	.
14	Of	1	49
		2	65
		3	73

Why^are^NP-complete^^problems^^interesting?^First,^although^no^efficient^algorit
hm^for^an^NP-complete^problem^has^ever^been^found,^nobody^has^ever^proven^
that^an^efficient^algorithm^for^one^cannot^exist.^In^other^words,^it^is^unknown^
whether^or^not^efficient^algorithms^exist^for^NP-complete^problems.^Second,^the
^set^of^NP-complete^problems^has^the^remarkable^property^that^if^an^efficient^al
gorithm^exists^for^any^one^of^them,^then^efficient^^algorithms^exist^for^all^of^th
em.

Fig. 5. The stego-text

3.2.1 Extraction Phase

The receiver can extract the embedded data through reversible computing to get the
combination number d_i. We elaborate the extraction algorithm in the following.

Algorithm 2: Message Extraction by Multiple-Base Encoding

Input: a stego-text S.
Output: the secret data D.
Steps:

1. Sequentially scan the stego-text to construct the *Text linked list*. If there are two
   blanks after this word, then set 1 in the *Shifting bit*. Otherwise, write 0 in the *Shifting bit*.
2. Calculate the repeated words from the linked list to generate the *Correlation table*.
3. Calculate $b_{i\,(i=1,2,...,n)} = |W_{i\,(i=1,2,...,n)}|$ respectively according to the *Correlation table*.
4. Scan the *Text linked list*: If the *Shifting bit* equals 1, then record the corresponding
   Occurrence number W_{ij} of the *Correlation table*. Then set d_i is equal to the associated W_j , where $i=1, 2,...,n$.
5. Compute an integer $x = d_1 + \sum_{i=2}^{n}(d_i \cdot \prod_{j=1}^{i} b_j)$ and then derive the secret data D using

   binary transformation.

Finally, the receiver decrypts D using the secret key K. Our scheme can avoid the drawback of Chen et al.'s scheme since we employ a linked list to adjust inter spaces instead
of pair words. Of course, in order to increase the hiding payload, our scheme can be
extended to a better method if we change the shift ways mapping to a binary equation.
It's embedding payload will equal to $\prod_{i=1}^{n} 2^{b_i}$. In addition, to make our scheme robust,
we can employ TeX tool or PDF formats to protect the inter spaces from the tampering
attack. Meanwhile, our scheme will not result in error of ending lines.

4 Experimental Results and Discussions

In this section, some experiments results are conducted and discussed to demonstrate
the performance of the proposed scheme. Herein, the evaluated criterion on the

proposed scheme mainly concerns on the hiding capacity, which refers to the amount of the hidden data. To confirm whether the proposed scheme is feasible for various text documents, in the experiment, we used three kinds of text documents including news, stories, and technical papers to be the test documents. More precisely, as regards the news, at first, we randomly choose three different news such as technology news, entertainment news, and health news as the test documents. Next, as to the stories, the test documents include science fictions, fairy tales, and classic short stories. Eventually, for the technical papers, papers for e-learning, information hiding, and distance education are as the test documents. In addition, it is worth noting that each test document only consists of 5000 words for the reason of a fair comparison.

Table 2 shows the comparison in embedding capacity for various test documents. As we can observe in the table, the fact reveals that the proposed scheme is flexible for a variety of documents. The test documents for classic short stories have a lower capacity than others because most words only appear once in the document. However, the hiding capacity of our proposed scheme can be satisfied and applied in most documents.

Table 2. Comparisons of the hiding capacities

Test document (5,000 words)		Hiding capacity (bits)	Average capacity (bits)
News	Technology news	1079	1076
	Entertainment news	1089	
	Health news	1061	
Stories	Science fictions	1000	1000
	Fairy tales	1080	
	Classic short stories	920	
Technical papers	E-learning papers	1065	1103
	Information hiding papers	1125	
	Distance education papers	1119	

5 Conclusions

In this paper, a text hiding method has been proposed. The proposed scheme can fulfill invisible security, while keeping computation load in $O(n)$. Furthermore, to validate the hiding payload, we evaluated 3 kinds including 9 categories of documents. The results show that the proposed scheme is available and flexible for a variety of documents. Especially, our scheme can eliminate side effects of Chen et al.'s.

References

1. Kutter, M., Winkler, S.: A Vision-Based Masking Model for Spread-Spectrum Image Watermarking. IEEE Trans. Image Processing 11, 16–25 (2002)
2. Cox, I.J., Kilian, J., Leighton, F.T., Shamoon, T.: Secure Spread Spectrum Watermarking for Multimedia. IEEE Trans. Image Processing 6(12), 1673–1687 (1997)
3. Chang, C.C., Tsai, P.Y., Lin, M.H.: SVD-based Digital Image Watermarking Scheme. Pattern Recognition Letters 26(10), 1577–1586 (2005)
4. Chang, C.C., Hu, Y.S., Lu, T.C.: A Watermarking-Based Image Ownership and Tampering Authentication Scheme. Pattern Recognition Letters 27, 439–446 (2006)
5. Chang, C.C., Lu, T.C.: Lossless Information Hiding Scheme Based on Neighboring Correlation. International Journal of Signal Processing, Image Processing, and Pattern Recognition 2(1), 49–56 (2009)
6. Chang, C.C., Kieu, T.D., Wu, W.C.: A Lossless Data Embedding Technique by Joint Neighboring Coding. Pattern Recognition 42, 1597–1603 (2009)
7. Tai, W.L., Yeh, C.M., Chang, C.C.: Reversible Data Hiding Based on Histogram Modification of Pixel Differences. IEEE Trans. on Circuits and Systems for Video Technology 19(6), 906–910 (2009)
8. Chang, C.C., Lin, C.C., Tseng, C.S., Tai, W.L.: Reversible Hiding in DCT-Based Compressed Images. Information Sciences 177, 2768–2786 (2007)
9. Wang, K., Lavoue, G., Denis, F., Baskurt, A.: Hierarchical Watermarking of Semiregular Meshes Based on Wavelet Transform. IEEE Transactions on Information Forensics and Security 3(4), 620–634 (2008)
10. Bender, W., Gruhl, D., Morimoto, N., Lu, A.: Techniques for data hiding. IBM Systems Journal Archives 35, 313–336 (1996)
11. Wu, C.C., Chang, C.C., Yang, S.R.: An Efficient Fragile Watermarking for Web Pages Tamper-Proof. In: Chang, K.C.-C., Wang, W., Chen, L., Ellis, C.A., Hsu, C.-H., Tsoi, A.C., Wang, H. (eds.) APWeb/WAIM 2007. LNCS, vol. 4537, pp. 654–663. Springer, Heidelberg (2007)
12. Liu, T.Y., Tsai, W.H.: A New Steganographic Method for Data Hiding in Microsoft Word Documents by a Change Tracking Technique. IEEE Transactions on Information Forensics and Security 2(1), 24–30 (2007)
13. Hwang, M.S., Chang, C.C., Hwang, K.F.: A Watermarking Technique Based on One-way Hash Functions. IEEE Transactions on Consumer Electronics 45(2), 286–294 (1999)
14. Chen, C., Wang, S., Zhang, X.: Information Hiding in Text Using Typesetting Tools with Stego-Encoding. In: Proceedings of the First International Conference on Innovative Computing, Information and Control, Beijing, China, pp. 459–462 (2006)
15. Zhang, X.P., Wang, S.Z.: Steganography Using Multiple-Base Notational System and Human Vision Sensitivity. IEEE Signal Processing Letters 12(1), 67–70 (2005)
16. Cormen, T.H., Leiserson, C.E., Rivest, R.L., Stein, C.: Introduction to Algorithms, 2nd edn. The MIT Press, Cambridge (2001)

Author Index